DICTIONARY OF BIOTECHNOLOGY

DICTIONARY OF BIOTECHNOLOGY

Ajit K. Ghosh

2008

CYBER TECH PUBLICATIONS
4264/3, Ansari Road, Daryaganj, New Delhi-110002 (India)
Ph.: 011-23244078. Fax: 011-23280028
E-Mail: cyberpublicationsdelhi@yahoo.com

Dictionary of Biotechnology

First Edition 2008

Published by G.S. Rawat

For Cyber Tech Publications

4264/3, Ansari Road, Daryaganj. New Delhi - 110 002

E-mail : cyberpublicationsdelhi@yahoo.com

Phone : 011-23244078,

Fax : 011-23280028

Printed at

Amit Enterprises, Moujpur, Delhi. Ph. - 9810878184

Preface

This dictionary explains all the key terms in Bio-technology. Each entry begins with a clear definition and is usually followed by an explanation, if necessary. The more important the term, the more detailed the entry.

This dictionary has been compiled to familiarise you with the language and terminology of Biotechnology. Biotechnology contributes to such diverse areas food production, waste disposal, mining, and medicine, Although biotechnology has existed since ancient times, some of its most dramatic advances have come in more recent years. Modern achievements include the transferal of a specific gene from one organism to another the maintenance and growth of genetically uniform plant-and animal-cell cultures, called clones; and the fusing of different types of cells to produced beneficial medical products such as monoclonal antibodies. The fuller explanation of more important or challenging concepts helps in tackling homework or coursework assignments.

This dictionary is a useful resource for students of undergraduate and vocational courses. The general readers including students and teachers will also find this as most useful and informative.

—Editor

A

A. Abbreviation for an Angstrom, 10^{-4} of a micron: there are 10,000 Angstroms in a millimeter.

A+T/G+ C Ratio An expression of the relative amount of adenine thymine pairs to guanine-cytosine pairs in a molecule of DNA.

A-Amanitin An Amotoxin which, at low concentrations, specifically inhibits eukaryotic RNA Polymerase II; RNA polymerase III is inhibited at high concentrations.

A-chromosome Chromosome belonging to the normal chromosome complement.

A-DNA Right-handed double-helical DNA with apporoximately 11 residues per turn. Planes of base-pairs in the helix are tilted 20° away from perpendicular to the axis of the helix. Formed from B-DNA by dehydration.

A-protein (In Tobacco Mosaic Virus). A mixture of small oligomers and monomers of coat protein subunits which occur in equilibrium with the larger 'disc' aggregates under conditions of physiological pH and ionic strength; coat protein occurs mainly as A-protein under conditions of high pH and low ionic strength.

Abaxial The side or on the distant from the axis; as on the lower surface of a leaf.

ABC Immunoperoxidase Method An immuno-peroxidase method involving the use of a preformed avidin-biotin-peroxidase complex (ABC) which has surplus biotin-binding capacity. Initially, a ('primary') antiserum is raised against the required antigen: if the primary antiserum is derived from e.g. a rat, a 'secondary' anti-rat antiserum is prepared, and the anti-rat Ig antibodies are Biotinyl-ated. Tolocate a specific antigen, the section is treated with primary antiserum, washed, and hen treated with secondary antiserum; the subsequent addition of ABC localizes peroxidase at the site of specific antigen (since the ABC adheres non-specifically to biotin).

Abelson Murine Leukaemia Virus ((AbMuLV). A replication-defective, *v-one* Murine Leukaemia Virus isolated from a prednisolone-treated BALB/c mouse inoculated with Moloney murine leukaemia virus (MoMuLV). Ab-MuLV apparentlyarose by recombination between Mo-MuLV and mouse *c-abl* sequences; the *v-abl* product has tyrosine kinase activity. Ab-MuLV induces B-cell lymphoid keukaemia with a short latent period (3-4 weeks).

Aberation Disturbance of the rays of light so they are unable to be brought to a sharp focus.

Aberrant A growth that deviates from the normal or usual type; as exceptional growths (aberrations) which occur in some tissue cultures.

Abhymenial Of or pertaining to a region opposite or away from the Hymenum.

Abiogenesia (*spontaneous generation*) The spontaneous formation of living organisms from non-living material; apart from its application to the evolutionary origin or life, this doctrine has long been abandoned.

Ablastin Antibody which specifically inhibits reproduction of epimastigote forms of *try-panosoma lewisi* in the vertebrate host.

ABO Blood Group Major blood antigen system containing types A,B,AB, and O, depending on the presence or absence of the A and B antigens.

Aboral Away from or opposite to, the mouth, abortifacient Able to cause abortion.

Abort To stop or fail to reach full development (maturity) or to develop imperfectly; as some hybrids are prone to embryo abortion or as some attempts to culture a plant may fail (abortive attempt).

Abortion A miscarriage; arrested development of an organ

Abortive Transduction The transfer of bacterial genes into a new host by a viral vector that is not followed by integration of the new generic material into the genome of the recipient cell however, the transduced genes may persist for a time in the cell as a plasmid.

Abortive Transformation Temporary transformation of cell by a virus which falls to integrate into the host DNA

Absolute Containment Capacity for keeping or retaining completely an specified substance.

Absolute Plating Efficiency The percentage of cells that give rise to colonies. See *relative plating efficiency*.

Absolute Requirement A nutritional or environmental substance or condition that is necessary for plant growth and development.

Absorb To drink up or such, imbibe. The capacity for absorption (absorptive capacity) is inherent in some materials (absorbent) such as cotton wool (absorbent cotton). (2) To take in and utilize; as plants absorb water and nutrients through root hairs, or light utilizing it in photosynthesis.

Absorption Equipment, Floating Bed Scrubber A floating-bed-sieve-tray arrangement. Low-density spherical packing is kept on the sieve tray so that the gas bubbles rise through the wetter sphere. This system is advantageous for use with gas streams having particulate matter, because the fluidized character of the bed of sphere prevents the buildup of a particulate sludge.

Abyssal Zone That portion of the deep ocean where no light penetrates.

Acaricide A pesticide used to kill or control mites or ticks. Mites occasionally infest culture rooms and cultures, spreading pathogens. Acaricidal paint is applied to shelves and walls. Infected or old cultures must be discarded for mite control.

Acarospora A genus of Lichens (order Lecanorales). Thallus: crustose, areolate, with prominent areolae.

Acceptor Site The active site of an enzyme or a site on an organism, tissue or cell recognized by a biological molecule and with which it reacts in a highly specific manner.

Accession Plant or seed sample, strain or population held in a genebank or breeding programme for conservation and use.

Accessory Cells Those cells which, together with B Lymphocytes and/or T Lymphocytes, are involved in the expression of humoral and/or cell-mediated immune responses; they include e.g. Macrophages, dendritic Cells, and Langerhans' Cells.

Accessory Chromosome A supernumerary chromosome that is not necessary for the individual and not homologous to any of the normal chromosomes. (The term was used earlier to denote a sex chromosome.)

Accidental Evolution Evolution which confers no selective advantage and which owes its origin to mutations of more or les neutral effect.

Aclimate or Acclimatize To adapt or adjust physiologically to a new environment or climate. The process is acclimation or acclimatization and is regulated by natural processes or human cultural practices. Usage of acclimatize may be limited to adaptive responses involving changes in a selected environmental feature, such as light intensity.

Acclimation Adaption to climatic change on the part of the individual. The physiological adjustment or increased tolerance shown by an individual organism to a change in the surrounding environment.

Accretive Release A method of periodic introduction of biotic agents in which annual early season liberations against fairly abundant pest populations allow the beneficial organism population to increase naturally in response to rising pest densities as the season progresses.

Accumulator Plant A plant that absorbs certain elements which accumulate in its tissues to a much higher degree than in most plants, e.g., Equisetum concentrates large amounts of silica.

Acellular Refers to an organism, usually a protozoon, which consists essentially of a single cell but in which occur functionally specialized regions sometimes regarded as analogous to the organs and tissues of a differentiated multi-cellular organism. (2) Refers to an organism (e.g. a Virus) or structure (e.g. the stalk of Acytostelium) Which is not Cellular in any sense (3) Not divided into cells (as e.g. in a Plasmodium).

Acentric Fragment A fragment of chromosome without any centromere. Such chromosomes arise as a result of inversion, translocation, delection, Because of lac of centromere these chromosomes are lost during mitotic division.

Acentric Inversion An inversion of a segment of a chromosome which does not involve the centromere cf. Paracentric Inversion.

Acentric Applied to achromatid or a chromosome when it lacks a centromere. This condition may arise in an inversion heterozygote as a result of crossing-over between a normal and an inverted segment that does not include the centromere.

Acervulus A flat or saucer-shaped fungal Stroma supporting a mass of typically short and densely-packed conidiophores; acervuli commonly develop subcuticularly or subepidermally in a plant host, becoming erumpent at maturity, i.e., rupturing the overlying plant tissue to allow dispersal of the conidia, Some acervuli bear setae.

Acetification The aerobic conversion of ethanol to acetic acid by bacteria (usually Acetobacter spp). Ethanol is converted to hydrated acetaldehyde ($CH_2CH(OH)_2$) which is then dehydrogenated to give acetic acid. Acetification is an exothermic process.

Acetobacter A genus of Gram type-negative bacteria of the family Acetobacteraceae; the organisms occur e.g. on certain fruits and flowers, are responsible for some types of Beer Spoilage and Wine Spoilage, and are used e.g. in the manufacture of Vinegar. .

Acetobacterium A genus of Gram-negative, obligately anaerobic bacteria which occur in marine and freshwater sediments.

Acetoclastic Bacteria Organisms that form methane exclusively from acetic acid in anaerobic digestion. They show very low growth rats, with doubling times of several days, and often represent the rate-limiting step in biogas production.

Acetogen (1) Any bacterium – e.g. Acetobacterium woodli, Clostridium aceticum which can produce acetate, as the main product, from CO_2 and H_2 and/or from certain sugars. (2) (hydrogenogen proton-reducting acetogen) Any bacterium which can use protons as electron acceptors for the oxidation of certain substrates (e.g., ethanol, lactate, fatty acids) to acetate with concomitant formation of H_2.

Acetone/butanol Fermentation An industrial process, based on the fermentation of sugars (e.g., molasses), using Clostridium acetobutylicum to produce a mixture of acetone and butanol. This was the first fermentation to be carried out on a large scale using pure culture and aseptic methods. At one time, fermentation was the major source of these solvents. However, almost all plants were closed with the advent of the petrochemical industry. Interest in this process is now increasing as it may provide a means of producing solvents for addition to lead-free petroleum blends.

Acetosyringone One of the phenolic compounds released by wounded plant cells which can stimulate the transcription of the *vir* region of the *Ti* plasmid in the initial stages of transfer of DNA from *Agrobacterium* to the plant cell.

Acetylcholinesterase Refers to the "true cholinesterase" which catalysesthe hydrolysis of acetylcholine into choline and acetate. This enzyme is found in the central nervous system.

Acetylene Reduction Test A sensitive technique used to measure the ability of an organism to fix nitrogen. It is based on the fact that the enzyme responsible for the reduction of nitrogen gas (dinitrogen) is also capable of reducing acetyleneto ethylene. Material under investigation is incubated in a closed flask in an atmosphere of acetylene. The gas phaseis sampled periodically, and the constituent gases are separated by gas chromatography.

Acholeplasma A gamily of bacteria of the order Mycolasmatales; species of the sole genus. Acholeplasma, differ from the other members of the order e.g. in that their growth is not sterol-dependent.

Achromabacter An obsolete bacterial genus. For A. *anitratum*.

Achromat An objective lens in which chromatic aberration has been corrected for two colours (usually red and blue), and spherical aberration has been corrected for one colour (usually yellow-green).

Achromatic Figure The frame-work of 'fibres', or striations, which is formed between the two poles and the equatorial region, or between the two centrioles and the equatorial region, during cell division.

Achromatoplasm The non-staining protoplasmic reticulate cell-matrix, trickling filters, de-nitrification processes and anaerobic sludge digestion.

Achromogenic Refers to an organism (or e.g., ragent) which does not produce pigment (or colour); used e.g. of non-pigmented strains of normally Chromogenic organisms.

Acid Curd coagulation of casein in milk by an acid pH f 4.7 or lower.

Acid Fermentation One stage of anaerobic decomposition of sewage sludge, resulting in the biodegradation of complex organic compounds to simpler organic acids, as in anaerobic sludge digestion.

Acid Formers A group of facultative and obligate anaerobes, capable of hydrolysis, which ferment the complex organic compounds in sewage to organic acids, including acetic acid, CH_2COOH, and propionic acid, CH_3CH_2COOH. For effective anaerobic sludge digestion they should be in equilibrium with them ethanogenic bacteria. The main acid formers have been the obligate anaerobes, which are 10 to 100 times more numerous in digesting sludge than the facultative anaerobes.

Acid Phosphatase Enzyme with acidic pH optimum, which catalyses cleavage of inorganic phoshate from a variety of substrates. Found particularly in lysosomes and secretory vesicles.

Acid Secreting Cells Large specialized cells of the epithelial lining of the stomach (parietal or oxynticcells) which secrete 0.1N hydrochloric acid, by means of K^+/H^+ antiport ATPases on the luminal cell surface.

Acidaminococcus A genus of Gram-negative bacteria (family Veillonellaceae) which occur e.g. in the intestine in humans and pigs.

Acidic Amino Acids Amino acids having a net negative charge at neutral pH.

Acidic Having a pH of less than 7 (ahydrogenion concentration greater than 10^{-7} molar).

Acidophilus Milk A sour, medicinal beverage made by fermenting heat-treated, partially skimmed milk wit Lactobacillus acidophilus. (Viable L acidophilus appears to have a therapeutic effect on some intestinal disorders).

Aconitte Hydratase, Aconitase A hydratase which catalyses one stage of the tricarboxylic acid cycle, the reversible inter conversion of citrate and isocitrate.

Aconta Algae that do not have a flagellate stage in their life cycle. This is the characteristic of the red algae (Rhodohyta).

Acquired Character (1) Variation or modification of a character arising from the influences of environmental factors during the development of the organism and not by hereditary factors (2) Variously used for modification of phenotype or genotype, adaptive or non-adaptive, induced by one environment and not by another, (3) Theory of inheritonce of the acquired characters are inherited and adaptive, and thus account for evolutionary chage, (Lamarck, 1801).

Acquired Immune Deficiency Syndrume (AIDS) A virus-mediated disease with a very high mortality rat. AIDS was first applied to a narrowly defined disease thatwas clinically recognizable by symptoms of Kaposi'ssarcma and Pneumonocystis cornii pneumonia. However, it is now known to include several other life-threatening opportunistic infections. AIDS is a group of disease secondary to a defect in cell-mediated immunity associated with a single, newly discovered and unique virus. This virus has been described as human T-cell lymphotrophic-associated virus ad AIDS retrovirus The virus appears to be a retrovirus of the subfamily Lentivirinae which has previously been found in sheep, horses and goats. The spectrum of disease associated with and includes progressive encephalopathy (degeneraton of the brain). On infection, the virus infects a small number of mature T-helper lymphocytes as well as brain cells, where it is slowly replicated, in an interated and an unintegrated form. Both types of cells form infectious virions. Treatment is difficult since the integrated virus becomes part of the genome of brain cells. The virus kills those T-helper cells in which it replicates, causing mild or profound immune deficiency according to the number of cells destroyed. The blood and lasma of an infected individual are highly infectious, and the disease is spread through blood transfusions and preparations of blood clotting factors, as well as personal interaction. The virus shows antigenic drift. And antibodies when forced do not appear to affect the virus. Different isolates of the virus show a great variation in the ENV gene, which codes from the envelope glycoprotein, making it difficult to produce vaccines.

Acrasin In cellular slime moulds: a generic term for a chemotactic substance which is produced by cells and which serves as a chemoattractant for cell aggregation, acrasins are a diverse group of substances; they include camp in Dictyostelium discoideum (q.v.), a pterin in *Dictystelium lacteum*, and a dipeptide, 'glorin,' in *Polysphondylium violaceum*.

Acridine organge (basic organge; euchrysine; 3, 6-bis (dimethylamino)-acridinium chloride) A basic dye and Fluorochrome used e.g. in fluorescence Microscopy to distinguish between dsDNA (which fluoresces green) and ss nucleic acids (which fluoresces organge red). Sublethal concentrations of the dye are used for Curing plasmids.

Acridines Heterocyclic compounds which include acridine and its derivatives. At low concentrations, aminoacridines (e.g. roflavine (3,6-diaminoacridine), Quinacrine) appear to bind to dsDNA (or to double-stranded regions of ssDNA) primarily as intercalating Agents. At higher concentrations there is also a weaker, secondary type of binding in which the acridine binds to the outside of dsDNA or to ssDNA or ssRNA; the two types of binding may account for the differential staining of DNA and RNA by Acridine Orange. As *antimicrobial agents*, acridines are active against a wide range of bacteria.

Acroblast A body, or group of bodies, in the spermatid, derived from the substance of the idiozome and Golgiodies from which arises the acrosome, Variously called 'Idiosome', 'sphere', 'archoplasm,' etc.

Acronematic Refers to a eukaryotic flagellum which is smooth and tapers to a fine point.

Acropleurogenous Located both at the tip and on the sides of an elongated structure. Action A protein, found in most types of eukaryotic cell, which can polymerize (reversibly) to form non-contractile filaments (microfilaments) that are involved e.g. in maintaining ell shape and structure.

Actinobolina A genus of carnivorous ciliates (subclass Gymnostomatia).

Actinomadura A genus of bacteria (order Actinomycetales, wall type III; group: maduromycetes) which occur e.g. in soil; some species (A modurae, a. elletieri) can be pathogenic in man.

Actinomycin D (actinomycin C_1) An Antibiotic from Streptomyces sp; it contains a (red) substituted phenoxazone chromophore linked to two identical pentapeptide lactone rings.

Actinomycins Refer to a large group of peptide lactone antibiotics which are produced by various strains of Step-tomyces.

Actinorrhiza A bacterim-plant root association in which nitrogen-fixing root nodules are formed in certain non-leguminous angiospherms infected (through root hairs) by Frankia strains; the plants involved are typically woody pioneers of nutrient-poor soils in cold or temperate regions in the northern hemisphere.

Action Potential Animpulse in a neuron, propagated without change in magnitude, taking the form of a wave of deplolarization or reverse polarization imposed on a polarized cell surface.

Action Spectrum A graph of biological activity versus wavelength of light. It compares the effectiveness of light of different wave lengths.

Action, Biochemical Chemical changes resulting from the metabolism of living organisms.

Activating Enzyme An enzyme that couples a low-energy compound with ATP to yield a high-energy derivative. As used in this book, an enzyme that condenses ATP with a specific amino acid to give enzyme-bound AMP-amino acid for used in protein synthesis.

Activation Energy The energy barrier that bocks the tendency for a set of chemical substances to react. A reaction is speeded up if this energy barrier is lowered by providing a different reaction pathway with the aid of a catalyst. Designated by the symbol Ea.

Active Biomass The proportion of a cell culture that has microbiological activity.

Active Genepool Collection A collection of germplasm used for regerenation, multiplication, distribution, characterization and evaluation. Ideally, germplasm in the active collection should be maintained in sufficient quantity to be available on request. Active collection germplasm is commonly duplicated in a based collection and is often stored under medium to long-term storage.

Active Immunity Immunity that results from the production of antibodies within the body, rather than the introduction of antibodies by injection, for example, active immunity gives a high resistance against subsequent infection and may be developed as a result of the disease itself or by immunization (using killed or attenuated vaccines or toxoids). The immunity takes several days up to a couple of weeks to develop, lasts for a long time (possibly for life) and can easily be reactivated by a booster injection of the required antigen. Active immunization is used as a prophylactic.

Active Transport The transport of a substance across a biological membrane against a concentration gradient-that is, from a region of low concentration (of that substance) to a region of high concentration. Active transport requires the expenditure of energy and is a saturable process.

Adansonian Analysis Numerical type of analysis of the characteristics of an organism; each property of an organism (e.g., motility, sugar fermentations) is enumerated. Strains yielding the same or a similar total number are considered to be taxonomically closely related.

Adansonian Taxonomy A method of biological classification, proposed in the 18th century by Michel Adanson, in which relationships between organisms are defined by the number of characteristics which the organisms have in common; the same degree of importance ('weighting') is attached to each characteristic.

Adaptation (1) The process and the fact that individuals or populations of a species are "fitted" to their environment, (2) The process, and the result of the process, by which members of a genetic group, either as individuals or as parts of individuals or as group are fitted to past or present changes in their environment. Adaptation is understood to occur in form and function, and the environment to operate both ecologically and biologically.

Adaptive Enzyme An enzyme that is formed in response to an outside stimulus during adaptation. This term is now generally obsolete and has been replaced by *Inducible Enzyme*.

Adaptive Radiation The division of a single species into many species specialized to diverse ways of life.

Adaptive Value The proportion of the progeny of a given genotype surviving to maturity (relative to that of another genotype) is its adaptive value.

Adaptor Molecules Small RNA molecules (sRNA) that bind amino acis to their proper positions on an mRNA template during protein synthesis. Each is specific for both an amino acid and a template codon.

Adapter (1) A synthtic double stranded oilgonucleotide that is blunt ended at one end and at the other has a nucleotide extension that can base pair with a cohesive and created by cleavae.of a DNA molecule with a specific type Ii restriction endonuclease. After blunt-end ligation of the adaptor to the ends of a target DNA molecule, the construct can be cloned into a vector by using the cohesive ends of the adapter. (2) A synthetic single-stranded oligonucleotide that, after self-hybridization, produces a molecule with cohesive ends and an internal restriction endonuclease site. When the adaptor is inserted into a cloning vector by means o the cohesive ends, the internal sequence provides a new restriction endonuclease site.

ADCC (Antibody-dependent cell-medicated cytotoxicity). A phenomenon in which lgG-coated cells are killed by non-sensitised effect or cells (neutrophils, monocytes, NK cells) by a non-phagocytic mechanism which is n dependent of complement.

Additive Factors Cumulative factors, non-allelomorphic factors affecting the same character and enchancing each other's effect. Such factors are said to show Additive Effect.

Adenosine Phosphates, Adenine Ribonucleotides Important as components of nucleic acids and as the major form in which chemical free energy is stored and transferred. They are also important metabolic regulators for example in glycolysis and the tricarboxylic acid cycle. The biologically significant derivatives, including cyclic adenosine 3',5,-monophosphate, carry the phosphate ester on the C-5 of the ribose.

Adenosine Triphosphate (ATP) Compound in which energy is stored in high-energy phosphate bonds its components are the purine, adenine, D-ribose, and three phosphoric acid groups.

Adenylate Cyclase An enzyme which catalyses the synthesis of Cyclic Amp (cAMP) from ATP. In general, intracellular levels of camp are determined largely by regulation of adenylate cyclase activity and/or synthesis.

Adenylic Acid The ribonucleotide containing the purine adenine.

Adenylylsulfate Reductases Enzymes of sulfur metabolism which reduce either phosphoadenylylsulfate (APS reductase) or adenylylsulfate. Adenylylsulfate reductase, has been indentical with one component of the sulfate reductase in fulfate assimilation since adenylylsulfate is the donor of the sulfate group.

Adherent Anchorage Dependent Cells or cultures which will survive and maintain function only when attached to an inert substrate, e.g. plastic, glass, etc. The use of this term does not imply that the cells are normal nor that they are not neoplastically transformed.

Adler Test A test used for the identification of Leishmania spp. The organisms are cultured in an immune serum; in the presence of homologous antibodies, promastigotes develop in culusters or syncytia.

Adnexed Of the region of lamellastipe attachment in an agaric: extending for a length equal to only a small fraction of the depth of the lamella.

ADP-ribosylation The transfer of an ADP-ribosyl group from NAD+ to a protein, catalysed by an ADP-ribosyl transferase. In eukaryotic cells, various proteins may be ADP-ribosylated-apparently as anormal regulatory mechanism; certain bacterial toxins act by exerting an ADP-ribosyl transferase function; ADP-ribosylation also occurs e.g. in cells of Escherichia coli infected with Bacteriophage-4, the host RNA polymerase undergoing phase-induced ADP-ribosylation.

ADP (adenosine diphosphate) Unless otherwise specified is the nucleotide 5'ADP, adensonine bearing a diphosphate (pyrophosphate) group in ribose-O-phosphate ester linkage at position 5 of the ribose moiety. Adenosine 2'5' and 3'5' diphosphates also exist, the former as part of NADP and the latter in coenzyme A and mRNA.

Adrenosterone An androgenic substance first extracted by Reichstein from the suprarenal cortex.

Adsorb The adhesion of a liquid, gas or dissolved material to a solid surface, resulting in concentration of the adsorbate [the adsorbed material] to the adsorbent [the adsorbing agent, such as activated charcoal]. The process is adsorption.

Advanced Waste Treatment Any treatment method or process employed following biological treatment to increase the removal of pollution load, to remover substances that may be deleterious to receiving waters or the environment or to produce a high-quality effluent suitable for reuse in any specific manner or for discharge under critical conditions. The term tertiary treatment is commonly used to denote advanced waste treatment methods.

Adventitious Produced in an abnormal or unusual position, or at an unusual time of development or away from the natural habitat; as when plant organs [bus, shoots, roots, others] develop on callus or nonzyotic embryos [embryoids] develop without an ovary or fertilization on callus (adventive embryony].

Aerated Grit Chamber Several make of grit channel are aerated with coarse air bubbles for removing the separation of the grit from the organic solids. Aeration that takes place after grit removal is called pre-aerating because it immediately precedes the main sewage treatment.

Aeration Period The detention period in an aeration tank. From 6 to 8 h at dry-weather flow is typical for domestic settled sewage.

Aerobic Waste Treatment A biological process in which aerobic microorganisms are grown on wastewater as a means of lowering the biochemical oxygen demand. Rapidly growing organisms use carbon compounds as source of material for cell growth and metabolic energy, generating cell biomass and carbon dioxide.

Aerofilter A commercial term applied to a trickling filter containing a relatively coarse filtering material and operating at a high rate which may be maintained, if necessary, at times of low sewage flow by recirculation of the filter effluent or other diluting liquids.

Aerosols Vapors or particles suspended in air; chemical aerosols may be used to kill microorganisms. A colloid of liquid or solid particles suspended in a gas, usually air.

Aetilogy The study of causation. The aetiological agent of an infectious disease is the organism which initiates/causes the disease.

Affinity Chromatography Chromatography in which the immobile phase (bed material) has a specific biological affinity for the substance to be separated or isolated, such as the affinity or an antibody for its antigen, or an enzyme for a substrate analogue.

Affinity Labelling A technique used to identify the *Combining Site* of an antibody. In one method, the antibody is allowed to combine with a homologous *Hapten* (the' affinity lable') that carries an azide side-chain; on subsequent irradiation with ultraviolet light the azide forms a highly reactive nitrene group that combines with any of a range of organic groups in the combining site. The nature of the combining site may subsequently be determined e.g. by chemical analysis.

Aflatoxins A group of *Mycotoxins*, produced by strains of *Aspergillus flavus* and A. *parasiticus*, which contain a bifuran moiety fused at the 4,5 ring positions to a substituted coumarin. (cf. *Sterigmatocystin.*) Aflatoxins are soluble in organic solvents, and some exhibit fluorescence on UV irradiation; they are extremely head-stable. Aflatoxins are toxic to a wide range of exkaryotes; they can retard germination and growth in plants, inhibit the germination of several moulds (e.g. Mucor, Neurospora, Penicillium), and are hepatotoxic and hepatocarcinogenic in animals. Severe outbreaks of disease have occurred in domestic animals and poultry fed with aflatoxin contaminated feedstuffs-particularly groundnut and cereal feeds.

Agamic, Agamous Having no pistils, stamens nor true seeds; reproducing asexually by apomixis.

Agar Diffusion Test A Diffusion Test which differs from the Disc Diffusion Test in that, instead of employing antibiotic-impregnated discs, a solution of each antibiotic is allowed to diffuse from a separate 'well ' cut into the agar.

Agar or Agar-Agar A gelatinous polysaccharide obtained from the red algar Gelidium corneum and from several other red algar. It is a solidifying agent that mixed with nutrient media (0.6-1.0%). Forms a gel for growing tissue cultured plants and for other purposes. It ranges in quality from relatively inert to very impure (complex, undefined). Its firmness as a gelling agent is affected by medium pH (it is softer when the medium is more acidic) and salt concentration (it is softer when the medium is more dilute). Agar gels melt at about 100°C but solidity at about 44°C.

Ageing of Lakes Lakes may become marsh and eventually dry up due to the stimulation of their plant life by natural eutrophication. The addition of human wastes or farm fertilizers my bring about the ageing.

Agent Any power, principle, orsubstance capable of acting upon the organism, whether curative, morbific, or other.

Agglutinin A specific antibody in blood plasma serum, causing agglutination (clumping) of blood cells with an incompatible agglutinogen Agglutinins may occur naturally, or may develop as a response to the introduction of an agglutinogen.

Aging Cask A barrel-shaped vessel of staves, headings, and hoops usually of oak, used to store wine for a period during which taste and aroma develop before it is bottled.

Agouti The gray-brown coat colour of certain wild animal, Individual hairs are black with a subapical band of yellow near tip. The designation comes from an animal species of he same name.

Agrenet The computer communications network established by the Agricultnre and Food Research Council (AFRC) for the agriculture and botanical research community in the United Kingdom.

Agricultural Alcohol Ethanol that has been produced by fermentation of a raw material derived from agriculture. The main sources of agricultural alcohol are sugar cane, maize, grapes, molasses, potatoes, spoilt fruit, wheat, cassava and rice.

Agricultural Biotechnology Agricultural biotechnology is a technology using the sciences of molecular genetics, microbiology, physiology, biochemistry and biophysics for manipulating fundamental basic biological processes and improving the productive capabilities of the sol, crop plants and the farm animals. This definition thus makes Agricultural Biotechnology essentially a biological technology for agricultural product improvement Clearly, in Agricultural biotechnology, the applied disciplines of plant breeding, agronomy, horticulture, plant pathology, etc, are integrated with the sciences of molecular genetics, microbiology, physiology, biochemistry and biophysics in ways such that fundamental basic bioprocesses are manipulated for beneficial use with the result that new crop varieties or technologies with improved yield potential or crop quality become available for commercialization in agriculture viewed as a primary industry.

Agricultural Pollution The liquid and solid wastes from all types of farming including runoff from pesticides, fertilizers and feedlots; erosion and dust from plouging, animal manure and carcasses and crop residues and debris. It has been estimated that agricultural pollution in the United States amounted to more than 2½ billion tons per year.

Agricultural Waste Utilization A number of biological or fermentation processes are employed; the production of compost or silage; the production of methane by anaerobic digestion; the production of fuel alcohol (ethanol) and other solvents by yeast or bacterial fermentation; the production of glucose and fructose syrups using enzyme methods; the production of single cell protein, using fungi or bacteria, as an animal feed or human food.

Agrochemicals Crop-protection chemicals, which may be synthetic, semisynthetic or natural products, that are used to increase crop yields. Agrochemicals include pesticides, herbicides and fungicides.

Agroecosystem The modified and simplified system of plant, animals and habitat used for human agricultural purposes

Agroinfection A method for introducing into a plant infectious DNA (or a cDNA copy of RNA) from a plant virus: the DNA (more than one genome's length) is incorporated into the T-DNA portion of a Ti plasmid, the plasmid is introduced into *Agrobacterium tumefaciens*, and the plant is then infected with this organism the viral DNA is transferred to the plant cells with the T-DNA.

Air Sterilization The removal of contaminating organisms from air prior to its use as the source of oxygen and/or for mixing cultures. Air sterilization processes are required to remove organism with dimensions as small as 0.5 m.m. from very largevolumes of air. Heat generation during air compression is of value in killing some organisms, but in general it is necessary to filter the air using fibrous, sintere oracetylated polyvinyl alcohol sheet filters.

Airlift Fermenter A Loop Fermenter in which the circulation of the culture is typically achieved by pumping air in at the bottom of a Draft Tube (or at the base of the annulus) – the air being voided via an opening in the top of the column; the air bubbles lower the hydrostatic pressure of culture in the draft tube so that culture continually flows down the annulus an up into the draft tube. (Reversed flow occurs if air is bubbled into the annualus.) The ICI pressure-cycle fermenter (used for Pruteen production) is an example of a tubular loop airlift fermenter. It consist essentially of two tall vertical columns (of different diameters) which communicate at top and bottom, and there is no draft tube; air is pumped into the bottom of the wider column (the 'riser') and promotes circulation on the airlift principle. This fermenter achieves a high level of dissolved oxygen sine oxygen solubility is increased by the increased hydrostatic pressure at the base of the (tall) riser; additionally, the physical separation of riser and down comer facilitates the removal of heat by allowing the inclusion of a heat-exchanger in the down comer.

Albuminoid Nitrogen Analysis An obsolescent but quick analysis of sewage-determining the fraction of the combined organic nitrogen that is easily decomposed by alkalinepotassium permanganate, KMnO, after the release of any ammonia by distillation.

Alcaptonuria An inherited metabolic defect in man, characterized by the excretion in the urine of excessive amounts of homogentisic acid.

Alcohol Dehydrogenase (ADH) An enzyme catalyzing the conversion of alcohols into aldehydes and ketones, and the reverse.

Alcoholic Fermentation A type of Fermentation (sense 1). Carried out by various yeasts and other fungi (e.g. species of *Saccharomyces, Pichia, Aspergillus, Fusarium, Mucor*) and or certain bacteria (e.g., Zmomonoas), in which ethanol is formed by D-glucose (or certain other sugars, depending e.g. on organism). In e.g. Saccharomyces, glucose is converted to pyruvate via the Embden-Meyerhof-Parnas pathway, pyruvate is decarboxylated to acetaldehyde by pyruvate decarboxylase and thiamine pyrophosphate, and acetaldehyde is then reduced to ethanol by NAD-dependent alcohol dehydrogenase – thus allowing reoxidation of the NAD reduced during the EMP pathway. Small amounts of side-products are usually formed, e.g., Glycerol, acetaldehyde, lactic acid, 2,3-butanediol. Succinic and acetic acids, and Fusel Oil; these occur in proportions which depend on organism and conditions. Alcoholic fermentation by Saccharomyces spp is widely exploited commercially.

Alcohols Under appropriate conditions certain alcohols can be rapidly lethal to a range of bacteria, fungi and viruses; they have little or no effect on endospores.

Aleurone Layer (1) Outermost layer of endosperm in some seeds, rich in aleurone (proteinaceous) grains; typically triploid ells. (2) The external layer of cells in the endosperm of cereals. Aleurone grains are located in this single layer of cells. Algae A heterogeneous group of unicellular and multicellular eukaryotic photosynthetic organisms.

Algal Bloom (Phytoplankton bloom). A sudden proliferation of microscopic algae in water bodies, stimulated by the input of nutrients (e.g., phosphates).

Algal Diseases Algae are subject to various diseases of microbial aetiology, some of which are of economic importance in seaweeds cultivated for food etc. Thus, e.g., disease of Laminaria japanica ('habidai') include 'frond twist disease' caused by a mycoplasma-like organism, and various rots caused by alginate-degradig bacteria; sporelings in culture may be killed by H2S produced e.g. by sulphate-reducing bacteria Other microorganisms which can infect algae, but whose pathogenicity is uncertain, include e.g., *Labyrinthulas*, *Phagomyxa*, and members of the *Plasmodiophoromycetes*.

Algal Harvesting The growth of algae (e.g., in a lagoon of sewage effluent) so as to reduce its nitrogen and phosphorus concentrations. The algae are periodically removed.

Allelomorph (1) An old term that is synonymous with alleles. (2) the two states of a factor (gene) conditioning the expression of a pair of contrasting characters. The alleles occupy exactly these locusin both homologous chromosomes but the identical factors situated in two homologous chromosopmes (in a true breeding type) should not be called alleles. Allele is a short form ofallelomorph. (3) One of two or more dissimilar Factors of Genes which on account of their corresponding position in corresponding chromosomes are subject to alternative (mendelian) inheritance. Two factors producing unrelated physiological effects, but apparently ineparable, may be taken provisionally as nonallelomorphic, e.g, yellow and scute in *Drosophila melanogaster*.

Allerton Disease An African Cattle Disease which involves a mild febrile condition followed by the appearance of skin nodules; it is caused by the bovine mammillitis virus and is apparently similar or identical to pseudo-*Lumpy Skin Disease*.

Allometry The relation between the growth-rare of a part of an individual and the growth-rate of the whole or of another part; the relationship between growth-rate of the whole or of another part; the relationship between growth-rate of different groups, races, genera, etc.

Allomonas A genus of bacteria (family Vibrionaceae) which have been isolated from fresh water, sewage and faeces; GC%: ca. 57. Type species: A. enterico

Allomone A chemical substance produced or acquired by an organism, which, when it contacts an individual of another species in the natural context, evokes in the receiver a behavioral or physiological reactions adaptively favorable to the emitter; cf. kairomone.

Allomorphosis The relation of parts of organisms at some definite age to wholes or parts also at some definite age but of different groups (race, varieties, species, genera) e.g. egg size of hatching weight to adult size or weight.

Allopoyploid A polyploid whose chromosomes form at meisosis not multivalents, but, so far as their homologies permit, bivalents always in the same way, e.g., a tetraploid with 2n bivalents or a triploid with x bivalents and x univalents (Darlington and Mather (1950).

Allosteric Enzyme An enzyme that shows unusual kinetic properties in response to changes in substrate concentration and/or in response to the action of allosteric effectors that are not substrates. It the rate of reaction is plotted as a function of substrate concentration for such enzymes, the result will be sigmoid curve rather than the usually right angled hyperbola. Addition of the appropriate allosteric usual hyperbolic form.

Allosteric Transition A change from one conformation of a protein to another conformation.

Allotype Subclasses of immunoglublins which exhibit different amino acid sequences within their constant regions.

Allylamines A group of synthetic antifungal Agents which are highly active against dermatophytes and show somewhat variable activity against yeasts such as Candida spp. The group includes nafitifine, currently in clinical use as a topical antifungal agent. Allylamines apparently act by inhibiting the fungal enzyme squalene epoxidase.

Alpha Amylase An Enzyme naturally occurring in wheat seed at germination, causing starch breakdown to sugar. If unduly active in milling wheat it reduces the quality of bread and cakes. Tested for a by either the hagberg test or the farrand test.

Alpha Factor For an aerator in an activated sludge plant, the rate of oxygen transfer in clean water divided by the rate of transfer in mixed liquor.

Alphanochaete A genus of filamentous, prostrate, branched green algae (division chlorophyta) which grow as epiphytes on other freshwater algae. Some of the cells in the filaments bear unicellular 'airs' with bulbous bases. Quadriflagellate zoospores or aplanosores are formed; sexual reproduction is oogamous.

Alternate Dominance A theory of sex-determination which supposed al individuals to be heterozygous for sex but that the male determiners were dominant in male offspring and the female determiners dominant in female offspring.

Alternating Double Filtration (ADF) To increase the organic load on a trickling filter, and to avoid ponding, two filters can be placed in series, with a humus tank after each. When the primary filter shows signs of ponding, or after a set time (e.g. 1 week), the two filters are reversed, with the second filter operating before the first one. There is an additional pumping cost, but in a plant with several filters and humus tanks there may be no need to build extra humus tons. To obtain an effluent of BOD_5; SS quality 20:30 organic loads up to 0.25 kg at BOD_5, per m^3 can be used.

Alternation of Generations In the life cycles of some organisms; the alternating formation of one or more generations of mature haploid individuals and one or more generations of mature diploid individuals; in this context 'mature' refers to the ability of the organisms to produce reproductive cells (gamets or spores). An alternation of generations occurs e.g. in some fungi (e.g. *Allomyces* spp), in certain protozoa of the *Foraminiferida*, and in many algae (e.g., *Laminaria*).

Alternative (alternate) Complement Pathway Pathway in which the complement cascade is activated at the C3 level; immune system products are not required for the activation of this pathway.

Alternative Technology Technology that aims to use resources sparingly and to do the minimum damage to the environment or to species inhabiting it, while permitting the greatest possible degree of personal control over the technology on the part of its user.

Amatoxins Toxic cyclic peptides which occur in some species of *Amanita, e.g. A. phalloides, A. verna.*

Ambiguity Code A character other than the standard codes for representing the DNA and RNA bases (A. C.G.T and U) used to indicate a case with an ambiguous identity.

Ambiguity (1) The coding for more than one amino acid by a given codon. (2) the phenomenon that a given codon can be translated in more than one way. Used especially in relation to suppressible chain terminating mutations.

Ambisense RNA A viral ssRNA genome or genome segment which is positive-sense with respect to some genes but negative-sense with respect to others. Ambisense genome segments have been found in the Arenariridae and in the genus Phlebovirus. In these cases the –ve-sense sequences are transcribed directly into (viral-complementary) subgenomic mRNA; however, +ve-sense sequences are expressed only after genome replication, subgenomic mRNA being transcribed from the RNA strand complementary to the genomic strand.

Ambruticin An antifungal antibiotic (a cyclo-propylpolyene-pyran acid) obtained from a strain of Polyangium cellulosum; it shows in viro activity against e.g. *Candida* spp, dermatophytes, and other pathogenic fungi.

American Standard Code for Information Interchange ASCII A standard method of representing Alphanumeric characters in memory or on disk as singe integers in the range 0-127 (i.e. 7 bits).

Amino Acid A organic compound of the general formula $H_2N - CHR - COOH$, where R can be one of twenty or more different side groups. An amino acid is to named because it has both a basic amine group-NH_2 and an acdic carboxyl group, - COOH, Proteins are polymers of amino acids.

Amino Acyl Synthetase Anyone of at least twenty different enzymes that catalyze (i) the reaction of a specific amino acid with ATP to form amino acyl-AMP (activated amino acids) and pyrophosphate and (ii) the transfer of the activated amino acid to sRNA forming amino acyl-sRNA and free AMP.

Amino Sugars Monosaccharides in which an hydroxyl group has been replaced by an amino group has been replaced by an amino group ($-NH_2$). The amino group is often acetylated. Examples are D-galactosamine. D-glucosamine, D-mannosamine, neuraminic acid and muramic acid.

Aminoacyl-tRNA A tRNA molecule covalently bound to an amino acid via an acyl bond between the carboxyl group of the amino acid and the3-OH of the tRNA.

Amnoadipic Acid Pathway (AAA Pathway) A pathway forlysine biosynthesis which occurs only in certain lower fungi (Blastocladiales, chytridiles, Mucorales), ascomycetes (including e.g. Saccharomyces), basidiomycetes, and euglenoid flagellates.

Aminopetidases Expopeptidases, usually containing in metal ions, which shorten proteins and peptides from the N-terminal end of the chain, removing one amino acid residue per step.

Aminoquinolines A group of Antimalarial Agents which includes 4-aminoquinolines (e.g. chloroquine, amodiaquine) and 8-aminoquinolines (e.g. primaquine) – (chemotherapy). Chloroquine binds to DNA as an Intercalating Agent and inhibits DNA and RNA synthesis; however, its antimalarial action appears to be due to other effects.

Amitosis In certain eukaryotes: an atypical form of Mitosis (q.v.) in which the chromosomes do not condense, a spindle is not formed, the nuclear membrane persists throughout division, and the nucleus divides by constriction.

Amixis In certain haploid organisms: a type of reproduction, considered to be a deviant form of the sexual process., in which karyaogamy and meiosis do not occur even though morphologically differentiated structures (e.g. spore-containing asci) are formed; thus, e.g., the eight haploid ascospores formed (perascus) by Podospora arizonensis are said to be derived by mitotic divisions of each nucleus of the dikaryon in the ascus mother cell. (cf. *Apomixis*).

Ammonia Assimilation Ammonia, which can be used as the sole source of nitrogen by many microorganisms, may be obtained from an external source or produced intracellularly-e.g. by deamination reactions or by Assimilatory Nitrate Reduction. NH_2 can be taken up readily through the cytoplasmic membrane, but NH4+ uptake requires a transport system. Within the cell NH_3 may be assimilated in various ways.

Amphiaster The spindle and its two asters collectively.

Amphibian Toxins A group o chemically very heterogeneous toxins (biogenic amines, peptides, alkaloids, steroids) produced by toads, salamanders, frogs and newts, Pharmacologically, they include, heart, muscle and nerve toxins, sympathicomimetica, local anaesthetics and hallucinogens.

Amphidiploid (Amphiploid) (1) An allopolyploid. The result of hybridization between two plant species or genera where the chromosome set contributed

By each parent undergoes doubling and produces in the hybrid a chromosome number which is the sum of the diploid numbers of the two parent forms (2) A species or type derived from doubling the number of chromosomes found in the original F_1 hybrid of two different species: an allopolyploid.

Amphimictic Population A population with free crossing and vital and fertile descendants. Equivalent to panmictic population.

Amphora A genus of pennate Diatoms in which the cingulum is much wider on one side (the 'dorsa' side) than on the other, resulting in the two vales facing in more or less the same direction; the raphes in both valves can thus both make contact with the substratum, species are aquatic, benthic organisms capable of Gliding Motility.

Amplification Fragment Length Polymorphism (AFLP) A variant DNA amplification product produced by DAF, PCR, or RAPD of different size.

Ampulla (1) (Protozool.) In certain ciliates (e.g. some hymenostomes): a channel, or the wider part of a channel, leading into a contractile vacuole. (2) (Protozool.) In certain hypostome cilates: an organelle which has an adhesive function. (3) (mycol.) the expanded, terminal region of some types of conidiogenenus cell. (4) (mycol.) The expanded, terminal region of some types of conidiophore.

AMU (atomic mass unit). The basic unit of mass on a atomic scale, defined as 1/12 the mass of a carbon-12 atom. There are 6.023 x 10^{23} amu in one gram This number is known as Avagadro's number.

Amylase An enyme that hydrlyses the –1,4 glucosidic bonds of stach yielding dextrins and maltose.

Amyloid Protein A pathological, fibrillar, low-molecular-weight protein. In amyloidosis, it is deposited together wit glycoproteins and proteoglycans, primarily in the spleen, liver and kidneys.

Anabaena A genus of filamentous Cyanobacteria (section IV) in which the trichomes are made up of spherical, ovoid or cylindrical vegetative cells, heterocysts may be intercalary or terminal, GC%: 38-44. Species include e.g. A. azollae, A. cylindrical, A flos-aquae and A. variabills, distinguished e.g. on the basis of cell size and shape, filament form, etc; however, some of these characteristics may be variable in culture.

Anabolic Pertaining to an enzymatic reaction leading to the synthesis of a more complex biological molecule from a less complex one.

Anaboly An evolutionary change arising at the end of ontogenc. The adding, by a descendant, of a new stage on to the last stage of morphogenesis of the ancestral type; a modified form of overstepping. "Anaoly differ from 'overstepping' only that it is the final stage of morphogenesis instead of the definitive adult stage of the ancestor which is passed through in the descendant."

Anaerobic Biological Treatment Any treatment utilizing anaerobic organisms to reduce organic matter in wastes.

Anaerobic Jar A container used for the incubation of materials (e.g. inoculated media) in the absence of oxygen or, in general, under gaseous conditions other than atmospheric.

Anaerobic Reactor A fermenter or bioreactor that is run in the absence of air or oxygen.

Anaerobic Respiration Respiration under anaerobic conditions, the terminal electron acceptor being (according to species and/or environmental conditions): fumarate, nitrate, nitrite, nitrous oxide, sulphur (S°), sulphate etc.

Anaerobic Living or active onlyinthe absence of oxygen, or air, dissolved in th emedium (e.g. watr) or othewrwise available. Aaerobic bacteria can obtain the medium (e.g. water) or othwerwise available. Anaerobic bacteria can obtain oxygen from nitrates or sulphates, releasing nitrogen, nitrous oxide, ammonium or hydrogen sulphide, provided oxidisable material (e.g. ammonium or hydrogen sulphide, provided oxidisable material (e.g., vegetable waste, methanol, or other carbon compound) is available.

Anaerovibrio A genus of Gram-negatie bacteria (family Bcteroidaceae) which occur e.g. in the Rumen, Cells: curved rods, ca 0.5 x 1.2-3.6 um; monotrichously (polarly) flagellated. Carbon sourcesinclude e.g. fructose, lactate an ribose; major fermentation products from these substrates include propionic and acetic acids and CO_2. Fermentation of glycerol yields mainly propionic acid. Liplolytic, producing clearignon linseed oil agar. Type (only) species: A. Lipolytica.

Anagenesis The evolutionary change of a single lineage in the course of time

Anamnestic Response Recall phenomenon, or secondary response to an antigen with very rapid production of antibodies. The recall or the "remembering' by the immune system of a prior response to a given antigen.

Anaphase (1) Strasburger (1884) originally introduced this for the stage of nuclear division when the content of the nuclei were going back (Gk. Ana) to their normal appearance, but from about 1905 he used the term in the now universally adopted sense of the stage of mitosis or of meiosis I or II when the aughter – Chromosomes (or homologous chromosomes at meiosis I) move towards opposite poles of the spindle (2) A stage in cell division following metaphase in which the daughter chromosomes pass from the equatorial plate in metaphase midway to the two poles at the opposite ends of the cell. The anaphase is actually part of a continuous process of cell division, interrupted in its progress by being killed and stained. (3) The third stage of mitosis or of meiosis (or II) during which the chromosomes migrate toward opposite poles of the cell.

Anaptychia A genus of foliose or fruticose Lichens (order Lecanorales). Photobiont: a green alga. Apothecia: lecanorine: ascospores; brown, one-septate. Species may occur on e.g. rocks or bark.

Anaschistic Bivalents that are said to split "longitudinally" at the first mitotic division: these are bivalents with chiasmata close to the spindle attachment (as opposed to diaschistic).

Anastomosis The fusion of fungal hyphae, which results in cytoplasmic fusion (plasmogamy). It does not inevitably lead to immediate fusion of nuclei (karyogany).

Ancalomicrobium A genus of chemoorganotrophic, facultatively anaerobic Prosthecate Boteria found in aquatic habitats In high nutrient concentrations the cells are non-motile, rounded to rod-shaped; I low nutrient concentrations they first become ovoid, knobbed, subpolarly flagellate cells, sub-sequently developing into non-motile forms bearing two to eight cylindrical (occasionally bifurcated) prosthecae. Cell division occurs by budding from the cell body, One species: adetum.

Anchorage-dependent Cell A cell that will only grow and multiply if in contact with a suitablele oid support. Many animal call lines are anchorage-dependent.

Androhermaphrodite A plant in whose flowers the male organs are developed more strongly than the female; towards the close of the flowering seasons such a plant may show only male characteristics.

Andromonoecy The condition of an individual plant or of a species having individuals which bear both male (i.e. female sterile) and hermaphrodite flowers.

Androphage (male-specific phage). Any Bacteriophage which infects only host cells that contain a cnjugative plamid. Androphages include members of the genus inovirus and of the family leviviridae, (cf. Female-Specific Phage.)

Androsomes Chromosomes which occur only in nuclei of the male germline., never in somatic nuclei of either sex and never in the nuclei of the female germ line; male; limited chromosomes.

Anegy The absence of a hypersensitivity reaction (particularly a Delayed Hypersensitivity reaction) in a Primed subject exposed to the relevant antigen; for example, an anergic response to a tuberculin test is given by some tuberculous individuals-particularly those with extensive pulmonary tuberculosis.

Ang-kak (ang-quac, angkk; red rice). A chinese product (used mainly as a colouring agent for foods) made by the fermendtation of rice by monascus purpureus strains; the fungus produces two pigments; monascorubrin (red) and monascoflavin yellow).

Animal Holding Room Room meeting suitable isolation criteria used to house animals in cages before and after experimental use; may be in noncontaminated or contaminated areas.

Animal Viruses Viruses which can infect and replicate in the cells of animals-including mammals, birds, poikilothermic vertebrates and/or invertebrates.

Animalcules (Animaliculae). An archaic term for microscopic organisms, particularly protozoa.

Anisogamy The union of gametes which differ in form and/or physiology. When the gametes differ in size, the larger (macrogamete) is regarded as female, the smaller (microgmete) as male.

Anisokinetic Sampling The taking of air samples under conditions in which the velocity and direction of the air ambient air. When sampling velocity is greater than that of the ambient air, smaller particles are selectively collected over larger particles, and vice versa, resulting in counting errors proportional to the particle size and distribution.

Anisotropic Inhibitor A type of compound (e.g. monoazide ethidium) which inhibits energy transduction in Oxidative Phosphorylation by binding to certain negatively charged sites on the outer surface of the mitochondrial inner membrane.

Anealing The process of heating (denaturing step) and slowly cooling (renaturing step) double stranded DNA to allow the formation of hybrid DNA or DNA-RNA molecules.

Annelate Ascus An ascus whose wall has a small annular, apical thickening that is often amyloid or light-refractive; at maturity, the ascospores are released via a pore in the centre of the thickening.

Annulus Any of various structures which are ring-shaped (either actually or in cross-section). Examples include e.g. the outermost layer (= 'sheath') of the contractile stalk in some peritirich ciliates, the Periseptal Annulus, and the annular remnant of the *Partial Veil.*

Anoxygenic Refers primarily to photosynthetic processes in which oxygen is not present or produced.

Ansamycins A group of structurally related Antibiotics consisting of a macrocyclic ring which comprises an aromatic moiety spanned by an aliphatic bridge. Ansamycins (which include e.g. rifamycins, streptovaricins and tolypomycins) are produced by actinomycetes and appear to share a common mode of antibacterial action at high concentrations they also show activity against certain viruses in cell cultures.

Antagonism The inhibition of one entity by another of similar kind, Examples: one antibiotic may reduce or abolish the effects of another (e.g. a bacteriostatic drug will counteract the effects of a drug which acts only on growing cells); acid producing organisms may inhibit or kill acid-intolerant organisms sharing the same environment.

Anthracnose Any of various plant diseases-particularly those caused by fungi of the Melanconiales-in which discrete, dark-coloured, necrotic lesions develop on the leaves and/or fruits. For example, anthracnose of beans (Phaseolus spp) is caused by collectorichum lindemuthianum and is characterized by brown or black lesions (sometimes bearing masses of pink conidia) on the leaves, pods and seeds.

Anthracycline Antibiotics Antibiotics which have a tetrahydrotetracenequinone chromophore (containing three flat, coplanar, 6-membered rings). They function as intercalating Agents, inhibiting DNA and RNA synthesis. Daunomycin (=rubidomycin, daunorubin), produced by Strentomyces sp, is an anticancer agent which binds to dsDNA (ca. 12°) with no apparent base preference; it contains an aminosugar (daunosamine) which projects into the major groove an stabilizes the drug-DNA complex by electrostatic interaction with phosphate in the DNA. Adriamycin is very similar to daunomycin in structure and function. Other anthracyclines: nogalamycin (antitumour agent, ca. 18°), *cinerubin, rhodomycin.*

Anthrax Toxin A Plasmid-encoded Taxin, produced by Bacillus anthracis, comprising 3 protins: oedema factor (EF, factor I), Protective anting (PA, factor II) and lethal factor ILF< factor III); ach component alone has no biological activity, but, in the presence of PA, EF causes oedema in guinea pigs, and LF is rapidly lethal on injection into rats, PA possibly binds to eukaryotic cell surfaces, functioning as a receptor for EF and LF-which can then penetrate the cells. The biological effects of anthrax toxin seem to depend mainly on the (unknown) activities of LF, ef is an adenylate cyclease which is activated on contact with a eukaryotic cell component (?*Calmodulin*); the subsequent increase in camp causes oedma (PA is used as a vaccine for people at specil risk of *Anthrax*.)

Antiauxin A chemical that interferes with the auxin response. They may or may not involve prevention of auxin transport or movement in plants. Some said to promote morphogenesis in vitro; as 2,3,5-triiodobenzoate (TIBA mw 499.81), or 2,4,5-trichlorophenoxyacetate (2,4,5,-T mw 255.49), which stimulate the growth of some cultures.

Antibiotic (1) Originally, any microbial product which, in low concentrations (of the order of ug/ml), can inhibit or kill (susceptible) microorganisms. The term was used by Waksman in the 1940s to distinguish the (then) newly discovered drug penicillin (a product of the fungus Penicillium sp) from the synthetic (.e., man-made) sulphonamides which had been developed in the 1930s. Difficulties in terminology arose when the penicillins and other natural (i.e., microbial) antimicrobial products were modified-or synthesized de novo-in the laboratory. (2) currently, antibiotic is generally used to refer to natural, semi-synthetic and wholly synthetic antimicrobial compounds which are effective at low concentrations.

Many antibiotics are used as Chemother-apeutic Agents but others are too highly toxic for use and/or are ineffective within the humanor animal body; some are used for nontherapeutic purposes, and many or all antibiotics may have important ecological roles for therapeutic use an aantibiotic should, ideally, be toxic only for pathogenic micro-organisms-.e., it should be harmless to the host; however, many antibiotics can have adverse 'side effects', and some (e.g. penicillins) may provoke hypersensitivity reactions (e.g. Anophyloctic Shock) in certain patients. In many cases antibiotics are taken up via specific, often energy-requiring transport systems. Within the microorganism a given antibiotic chractertitically acts at specific target sites). (Particular antibiotics are cross-referred from entries dealing with their target sites: A given antibiotic may be Microbicidal or Microbistatic (some can be microbicidal or microbistiatic depending e.g. on conentration), and mixtures of antibiotics may exhibit antagonism or synergism. The antimicrobial activity of an antibiotic within the human or animal body may not correspond to its activity in laboratory tests, thus e.g. in vivo and antibiotic may bind to plasma proteins so that only a fraction of the administered dose is available for antimicrobal activity, (cf, e.g P-N succinyl sulphatriazole in Sulphonamides).

Microbial Resistance to Antibiotics Resistance to a given antibiotic ('drug resistance') is said to be constitutive (= intrinsic) in cells iwhich lack the specific target site(s) of the antibiotic and/or which are inherently impermeable to the antibiotic. A cell may acquired heritable resistance to antibiotic(s) in several ways. For example, the target site of a given antibiotic may be modified as a result of a chromosomal Mutation, such that it functions more or less normally nthe presence of otherwise inhibitory concntratins of that antibiotic a single mutant (resistant) cell in a population of susceptible cells can therefore grow and give rise to a population of resistant cells in the presence of the given antibiotic; in bacteria, chromosomal gene(s) specifying antibiotic resistance may be transferred to antibiotic-sensitive cells by *conjugation, Transduction* or *Transformation*, Resistance due to a single mutation is commonly effective against only one type of antibiotic-or against those antibiotics which share the target site. Alternatively, resistance to antibiotic(s) may result from the acquisition of an R Plasmid the acquisition of such a genetic element may confer on a cell resistance to a singe antibiotic (and often to related antibiotics-cf. Heteroresistance) or resistance to two or more unrelated antibiotics (multile drug *resistance*). R plasmid-or transposon-mediated resistance may involve e.g. the synthesis of antibiotic-modifying or antibiotic-destroying enzymes modification of the target site and antibiotic-effluxing system or acquisition of anantibiotic-insensitive target enzyme. In escherichia coli, phenotypic (non-heritable) resistance to certain antibiotics (e.g. chloramphenicol, ampicillin) can be induced by the presence of certain chemorepellents-e.g., acetylsalicylate (asprin), benzoate-which are detected specifically by the tsr gene product this phenomenon has been attributed to inhibition of OmpF porin synthesis by the chemorepellent. If a metabolic pathway is blocked b antibiotic activity, a cell may become insensitive to that antibiotic if it receives a supply of products of the blocked pathway.

Antibiotic-sensitivity Tests Laboratory tests are often carried out to determine the susceptibility of a pathogen to a range of antibiotics; the results of such tests may enable the clinician to select optimally active chemotherapeutic agents, and may be useful in identifying the pathogen.

Antibody (ab) Any Immunoglobulin molecule produced in direct response to an Antigen (or to an autocoupling Hapten) and which can combine specifically, non-convalently, and reversibly with the antigen which elicited its formation specificity is not necessarily absolute: (A Myeloma canproduce Ig with antibody

function in the absence of antigenic stimulation. The term 'antibody' is also used to refer to a homogeneous or heterogeneous population of antibodies. Antibody-antigen combination occurs by contact between specific antibody combining Sites and antigenic Determinants, and may promote complement Fixation, In vio, antibody-antigen combination may be beneficial if the antigen combination may be beneficial if the antigenic stimulus derives from an invading pathogen but can be harmful
Antibodies may be identified, assayed etc in vitro by a range of techniques *Gel Diffusion; Immunoassary; Immunoelec-Trophoresis; Immunofluorescence*; *Precipitin Test.* Detection and quantification of specific antibodies in samples of Serum are important in the diagnosis of many diseases.

Antibody Formation One form of Immune Response. When stimulated by a given Antigen, and by The cells (and/or their products), certain B Lymphocytes undergo Blast Transformation and proliferate; cells of the enlarged clone may differentiate to become Antibody-secreting Plasma Cells or may become memory cells, i.e., cells which have been 'primed' to react rapidly (in terms of antibody production) on a second or subsequent stimulation by the specific antigen. This constitutes the primary response. On subsequent stimulation by the same antigen the enlarged clone can respond, promptly, with an increased amount of specific antibody (the *secondary* or *anamnestic response*).

Anticomplementary (immunol) Refers to any substance e.g. Zymosan), or to any system (e.g. pre-formed antigen-antibody complexes) other than the specific combination of antibody with antigen, which brings about Complement fixation, if present in a Complement-Fixation Test, anticomplementary factors can-by fixing complement-simulate the effects of antigen-antibody combination, giving a false-positive reaction.

Antifungal Agents Chemical agents (including certain Antibiotics) which either kill or inhibit the growth of fungi (fungicidal and fungistatic action, respectively). (Antifungal agents are often called 'fungicides' regardless of whether their action is fungicidal or fungitatic.) Antifungal agents are used e.g. for the treatment or prevention of fungal disease of man, animals and plants, and as Preservatives for a wide range of materials. (a)

Medical and veterinary antifungal agents ('antimycotics') are used for the treatment of localand systemic mycoses such agents include e.g. Allylamines, Azole Antifungl Agents, Cycloheximide, Flucytosine, griseofulvin, Nitrofurans, Polyene Antibiotics, Salicylin Acid, tolnaftate. Many general Disinfectants and Antiseptics also have antifungal activity. (b) Agricultural antifungal agents (used against fungal diseases of crop plants) may function as Eradicants and/or as Protectants; they fall into two broad categories: contact (surface or non-systemic) antifungal agents, which adhere to plant surfaces but do not penetrate the plant tissues, and systemic antifungal agents, which are taken up by the plant and translocated within its tissues-usually in the transpiration stream (from roots to leaves). In general, a contact agent typically acts as a general enxyme inhibitor, while a systemic agent usually acts at a specific target site (process or structure) in the fungal cell. Fungi may rapidly acquire resistance to an agent which has a single target site, and continued use of such agents has led in many cases to the emergence and spread of resistant strains of fungal pathogens; the risk of emergence of resistant strains can be reduced by using mistures containing two or more agents which have different modes of action. Agricultural antifungal agents include e.g. Benzimidozoles (a), *Bisdithiocarbamates, Carboxin, Chloroneb, chloronitrobenzenes, copper sulphate Dicarbaimides, dichlofluanid, Dichlone, Dinitrophenols, dithiocarbamates, Dodine, Ethazole, Glyodin Hydroxypyrimidine Antifungal agents, Imazalil, Morpholine Antifungal Agents. Crtain Organophosphorus Compounds, Phenylamide Antifungal Agents, Phithalimide Antifungal Agents, Pachlraz, Pyracarbolid, Quinone Antifungal Agents, Sulphur, Thram, Triforine.*

Antigen (ag) Any agent which can elicit an Immune Response (cf. Hapten); 'antigen' may refer to an indivudla macromolecule or to a homogeneous or heterogeneous population of antigenic macromolecules. A given antigen usually contains more than one Determinant. An antigen may be soluble (e.g. microbial toxins, extracts) or particulate. The most effective antigens are proteins and polysaccharides in general, large molecules are more antigenic than are smaller ones, and antigenicity is enhanced by the presence of an Adjuvant, Minute quantities (e.g.

a few μg) of antigen may be sufficient to elicit amount required depending e.g. on the route of a administration. The surface of a microorganism typically consists of repeating patterns of antigens, and the classification of some groups of microorganisms is based on differences between the antigens of different strains.

Antigen-specific T Helper Cell A type of t helper cell which is activated by specific antigen. Most antigen specific T helper cells appear to be Carrier-specific T helper Cells; others, when stimulated by specific antigen bone on Antigen-Presenting Cells, release factors which act polyclonally n B cells-activating specific antigen-binding B cells without direct B cell-T cell contact.

Antioxidant (food sci.). A substane added to certain foods to inhibit oxidation of the lipids in those foods thus enhancing stability and prolonging shelf-life. Antioxdants currentlyused include e.g. ascorbic acid, butylated hydroxynaisole, butylated hydroxytluene, carotene, propyl-gallate and tertiary butvlhydroquinone. Some antioxidants have antimicrobial properties.

Antisense Construct A plasmid that contains the coding regions of gene placed adjacent to a promotor in such an orientation as to cause the gene to be transcribed from the DNA strand complementary to that from which RNA is normally transcribed.

Antisense DNA (1) The sequence of chromosomal DNA that is transcribed. (2) A DNA sequence that is complementary to all or part of a functional RAN (mRNA).

Antisense RNA In the original (narrow) sense: any ssRNA strand which, on hybridzing with complementary bases in an RNA or DNA molecule, inhibits the latters role if the 'target' is a transcript, the antisense RNA is counter transcript RNA (ctRNA), and if it is mRNA the antisense RNA is mRNA-interfering complementary RNA (micRNA). Currently, the 'antisense' concept broadly includes e.g. ssRNA-ssRNA interactions, ssRNA-dsDNA Interactions ('triple strand' or 'triplex'), ribozymes, and RNase H-mediated arrest; the feature common to these systems is that they are associated with nucleic acids in 'non-classical' roles. Antisense RNA regulates various in vivo systems: Antisense systems are being evaluated e.g. for antiviral activity.

Antistreptolysin O Test (asLT; ASO test; ASOT). A test used for detecting/quantifying serum antibodies to Streptolysin O; it is of use e.g. in the differential diagnosis of Rhenumatic Fever and rheumatoid arthritis. In one method, Inactivated Serum is serially diluted, and to each dilution is added a fixed volume of streptolysin O solution; after incubation at 4 °C for several hours (or 37 °C for ca. 15 minutes)-to allow antibody-streptolysin O combination-a fixed volume of RBC suspension is added to each dilution, and incubation carried out at 37 °C for 30-45 minutes. Complete absence of haemolysis in a given tube indicates that streptolysino in that tube has been fully neutralized by antibodies. The ASOT titre (often expressed in todd Units) is given by the highest haemolysis. ASOT can also be carried out as a Latex Particle test using streptolysin O-coated particles.

AOAC use-dilution Test A *Carrier Test* (q.v.) devised by the AOAC, which determines the Use-Dilution of a given disinfectant. To each of a number of dilutions of the disinfectant is added a (contaminated) small steel carrier; after immersion for 10 min, each carrier is transferred to broth to detect any surviving microorganisms, sporicidal activity is determined by exposing the endospores of Bacillus subtilis ATCC 19659 or Clostridium sporogenes ATCC 9081 to the disinfectant and subsequently testing for viability. Apanomyces A genus of predominantly aqutic fungi of the order Saprolegniales (q.v.) A species known as the 'MG fungus has been reported to cause a ycotic granuloma of fish while A. astaci causes Cryfish Plague Plant-pathgenic species include A. brassicae, A. Camptostylus, A. Cohilioides (n sugar-beet), a. euteiches (on peas) and A. raphani (on radish).

Appertization A method of Food Preservation in which certain types of food are exposed to a temperature/time regime which renders them safe for the consumer and microbiologically stable for extended, or indefinite, periods of time on subsequent storage in hermetically sealed containers at temperatures below 40 °C; appertised foods are not necessarily sterile, and in certain cases appertized foods must be kept under refrigeration. For foods which have no intrinsic properties (e.g. acidity) inhibitory to microbial growth and/or toxin production, appertization involves treatment which at least, destroys the spores of pathogenic bacteria (particular clostridium botulinum); acidic foods (below ca. pH 4.5), and those containing appropriate concentrations of e.g. sodium nitrite, are appertised at somewhat lower temperatures. High temperature short time (HTST) appertisation involves heat processing at high temperatures for periods ranging from ca. 2 sec to several minutes (cf. HTST Pasteurization). In the ultra-high temperature (UHT) appertization of milk the process involves e.g. maintenance of a minimum temperature of ca. 140 °C for atleast 2 sec.

Appressorium A specialized, flattened region of a hypha or germ tube formed by some plant-pathogenic fungi (e.g., Puccinia graminis, Venturia inacqualis); it aids infection by adhering closely to the surface of a host cell and producing a small outgrowth of fungal tissue (infection peg) which penetrates the host cell wall. A plant may respond to an infection peg by depositing a small region of wall-like material (a papilla) near the site of attempted penetration; such a papilla can sometimes penetration; such a papilla can sometimes prevent infection.

APT Paper Biotting paper impregnated with 2-aminophenylthioether before use, APT is chemically modified to its reactive diazo derivative, and is then able to bind (covalently) ssDNA, RNA and/or protein via the diazonium group.

Ara Operon (arabinose operon) An Operon concerned with themetabolism of Arabinose; when used without qualification, the term 'ara operon' usually refers to the araBAD operon, but in e.g. Escherichia coli three distinct genetic loci are involved in the uptake and metabolism of L-arabinose: araBAD (at ca. 1 minon the chromosome map), araFG (at ca. 45 min) and area (at ca. 61 min). The araBAD operon encodes enzymes for the conversion of L-arabinose to D-xylulose 5-phoshate; the araA gene encodes L-arabiose ketol-isomerase, araB encodes L-ribulokinase, araD encodes ribulose 5-phosphate 4-epimerase. The araFG operon specifies a high-affinity Binding protein-protein, araG the membrane omponent. The raE gene specifies

a protein involved in an independent low-affinity transport system.
All three operons are controlled by the the produc of th eregulator gene araC which is located next to araBAD but transcribed in th eopposite direction. In the absence of arabinose, the AraC protein behaves as a repressor of the araBAD operon, binding to the Operator and preventing transcription (negative promoter control. However, when arabinose is present it converts the AraC protein into an activator which binds to an initiator (aral) to promote araBAD transcription (positive promoter control). The araC gene is itself subject to autogenous regulation, i.e, the AraC protein represses its own synthesis.) Thus, expression o the AraBAD operon dependson the relative amounts of repressor and activator, which in turn depend on the level of arabinose (the co-activator). The *ara* regulon is also subject to Catalbolite Repression.

Arabitol A *Pentitol* formed e.g. by the reduction of *Arabinose.* It occurs, often with *Mannitol*, in various basidiomycetes and ascomycetes.

Arachidonic Acid In mammals: an essential C_{20} straight-chain fatty acid-$CH_3(CH_2)_3$ $(CH_2CH = CH)_4$ $(CH_2)_3$ COOH; it occurs e.g. as a component of cell membrane phospholipids, and (bound to protein) in blood. Free arachidonic acid is formed e.g. during the activation of lymphocyts, and is a precursor of e.g. Leukotrienes, Prostacyclin, Prostaglandins and thromboxanes.

Arachinia A genus of anaerobic (but aerotolerant), catalase-negtive, asporogenous bacteria (order Actinomycetales, wall type I). The organisms grow as a substrate mycelium which fragments into branched filaments and irregularly-shaped rods; aerial hyphae are not formed. Metabolism: fermentative, fermentation of glucose yielding primarily acetic and propionic acids; growth occurs readily in *vitro* on rich media at 37 °C, A. *propionica*, the type species (GC%: 63-65), is pathogenic in man and other animals; it is a common causal agent of *lachrymal Canaliculitis.*

Archaebacteria A phylogenetic category (kingdom) of prokaryotic organisms (cf. Eubacteria: Urkingdom.) Archaebacteria differ from both eubacteria and eukaryotes e.g. in the base sequences of their 16S rRNA and in that the archaebacterial cytoplasmic membrane contains branched, isoprenoid, either-linked lipids rather than fatty acid ester-lined lipids. In common with many eukaryotic genes, certain tRNA genes in some archaebacteria have been shown to contain introns. Other differences between archaebacteria and eubacteria include e.g. the structure of the DNA-dependent RNA Polymerase the nature of the cell wall (the archaebacterial Cell Wall does not contain Peptidoglycan: cf. Pseudomurein); and sensitivity to various antibiotics.

Arginine Dihydrolase (ADH) An enzyme system which catalyses the catabolism of arginine, with aconcomitant substrate-level phosphorylation; the system occurs in a range of bacteria. (1) *Arginic deiminase* hydrolyses arginine to citrulline and ammonia. (ii) *Ornithine carbamoyltransferase* catalyses a phosphorolytic cleavage of citrulline to form carbamoyl phoshate and ornithine. (iii) Carbamate kinase transfers the phosphate group of carbamoyl phosphate to ADP to form ATP and carbamic acid, the latter dissociating spontaneously to CO_2 and NH_3.

Armilaria ('Armillariella') A genus of mainly lignicolous fungi (Agaricales, Tricholomataceae). A mellea (the 'honey fungus') grows saprotrophically on many types of wood, forms a symbiotic association with certain orchids, and is parasitic on a range of deciduous and coniferous tree, on certain shrubs, and on some herbaceous plants. The fruiting body is highly variable in colour and appearance, the pileus being convex to flat (ca. 3-15 cm diam.), ochre to brown, with darker scales particularly near the centre: the upper part of the stipe, which often bears a wide annulus, is initially whitish, later reddish-brown. Basidispores ca. 8-9 x 5-6 μm. The fruiting bodies (which exhibit Bioluminescence) typically occur in clusters. The organism spreads by means of touch, black Rhizomorpns ('boot laces') which may be found e.g. under the bark of infected trees or in soil.

Aromatic Amino Acid Biosynthesis Phenylalanine (Phe), tyrosine (Tyr) and tryptophan (trp) are synthesized via the shikimate pathway; appendix IV (f) Although most of the steps in this pathway are apparently more or less universal, the control mechanisms and physical organisation of the enzymes (e.g., the formation of multienzyme complexes) vary between species and may be useful taxonomic criteria. The reactions by which prephenate is converted to Phe and Tyr also differ in different species. Thus, e.g., Escherichia coli and Bacillus subtilis synthesize Tyr via 4-hydroxyphenylpyruvate (HPP), and Phe via phenylpyruvate (PPy). Certain cyanobacteria and 'coryne forms' (*Corynebacterium glutamicu, Brevibacteriu* spp) synthesize Phe via PPy but lack prephenate dehydrogenase ad synthesize Tyr via arogenate (= 'pretyrosine'). Pseudomonas diminuta synthesizes Phevia arogenate but Tyr via HPP, white P. aeruginosa cansynthesize Phe and Tyr via PPy and Hpp, respectively and alsovia arogenate. Euglena gracilis synthesizes both Phe and Tyr via arogenate only.

Arterivirus A genus of viruses (family Togaviridae) which currently includes only equine arteritis virus. This virus infects equines, causing necrosis of the small arteries with various clinical manifestations (e.g. rhinitis, oedema, enteritis, bronchopneumonia, etc) Infection of pregnant mares commonly results in abortion. Transmission occurs horizontally and vertically; vectors are unknown. The virus an infect arrange of vertebrate cells in vitro.

Arthoniales An order of fungi of the Ascamycotina; members include crustose Lichens (photobiont green, often trentepohlioid) and lichenicolous and saprotrophic fungi. Ascocars are Apothecioid and my be lirelliform, irregular, etc; they containparaphysoids. Asci are bitunicate, usually clavate Genera: e.g. Arthonia.

Arthritis Inflammation of one or more joints. It may occur as a symptom or complication of various infectious disease-e.g. brucellosis, gonorrhoea, Haverhill fever, Lyme disease, rickttsioses, syphilis, tuberculosis, yaws; it may also be caused by certain viruses-e.g., parvovirus Ross River virus rubella virus, septic orthritis is due to infection of the synovial fluidcommonly be staphylococci e.g. secondarily to Osteomyelitis in an adjacent

Arthrospores A sexual spores formed by fragmentation of a filament in certain fungi.

Arthus Reaction Reaction resulting from injection of an antigen into the skin; inflammation results from activation of complement by IgG-antigen complexes.

Artificial Active Immunity Resistance to diseases which develops after administration of a vaccine or inactivated toxin.

Ascorbic Acid or Vitamin C [$C_6H_8O_8$ mw 176.12) A water soluble vitamin present naturally in some plants and also synthetically produced. Aside from its use as a vitamin, it is used as an antioxidant in plant tissue culture; included in disinfection solutions and sometimes in media.

Asepsis A condition in which there are no infections or contaminating microorganisms; the absence of sepsis.

Aseptic Technique The performance of a procedure or operation in a manner that prevents the introduction of septic material.

Aseptic Free of microorganisms that can cause contamination of infection.

Asexual Spores Spores generated by one cell, without fusion of two cells.

Asynchronous Communications A method of managing a dialogue between two devices such that data may be exchanged in an asynchronous manner (e.g. the RS22C protocol). Data may flow in both directions at once and each device can halt the flow from the other at any time in the dialogue using special flow control signals embedded in the data stream or appearing on special control lines.

Atmosphere The gaseous mass surrounding our planet. Also; a unit of pressure, equal to the normal pressure of air at sea level.

Atom The smallest unit of a chemical element. Consists of a nucleus and one or more electrons.

Atopic Hypersensitivity Immediate-type, lgE-mediaed common clinical allergies for which there is a geneticallydetermined predisposition forformation.

Atopy The chemical manifestation of type 1 hypersensitivity reactions including eozema, asthma and rhinitis. Where there is genetic predisposition.

Attenuation Change; used in reference to certain vaccines in which the organism or its products are reduced in virulence.

Autoimmune Disease Due to antibodies produced against the individual's own tissues; failure to differentiate between self and non-self.

Autonomic Nervous System The system (which in vertebrates comprises sympathetic and parasympathetic sub-systems) that controls such involuntary functions as those of guts and glands.

Authonomous Replicating Sequence (ARS) Any cloned DNA cloned DNA sequence that initiates and supports extra chromosomal replication of a DNA molecule in a host cell; often used in yeast cells. Also called autonomously replicating segment autonomously replicating sequence.

Autopolyploid A polyploid constructed entirely of sets of chromosomes that originated from the same species.

Autoradiography The detection of a radioactive substance in a cell or organism by putting it in contact with a photographic emulsion and allowing the material to "take its own picture." The emulsion is developed, and the location of the radioactivity in the cell is seem by the presence of silver grains in the emulsion.

Auxin-cytokinin Ratio The relative proportion of auxinto cytkinin present in plant tissue culture medium. Varying the relative amounts of these two hormone groups in tissue culture formals affects the proportional growth of shoots and roots in virto. As the ratio is increased [increased auxin or decreased cytokining contact]. Roots are more likely to be produced, an as it is decreased root growth declines and shoot initiation and growth are promoted. This relationship was first recognized by C.O. Miller and F. Skoog in the 1950's.

Axenic Culture A culture without foreign or undesired life forms. An axenic culture may include the purposeful cultivation of different types of cells, tissues or organism.

Axenic A pure culture of one species. This implies that cultures are free of micro-organisms (aseptic or germ free).

Axillary Bud Proliferation Propagation in culture by protocol and media which promotes axillary (lateral shoot) growth. This is a technique for mass production (micropropagation) of planlets in culture, achieved primarily through hormonal inhibition of apical dominance and stimulation of lateral branching.

B

5-Bromodeoxyurdine Growing cells readily incorporate this base analog into their nucleic acids and can be killed by light exposure. It suppresses the utilization of deoxythymidine. This procedure is a selection mechanism utilized in mutaginic studies to eliminate wild type cells and recover the auxotrophs.

B Cell A major population of lymphocytes; the B cell stem cells are found in bone marrow (in the bursa of Fabricius in avian species); mature B lymphocytes have cell-surface lg receptors for specific antigenic determinants; antien-stimulated B cells proliferate and differentiate into antibody-producing plasma cells.

B-chromosome A special kind of chromosome which does not contain any major gene, and hence are genetically almost inert. However, their presence reduces the fertility of the plant. They are usually heterchrmatic but may contain centromere. During division the B-chromosomes are distributed to two poles in unequal numbers.

B2-microglobulin A polypeptide which constitutes part of some membrane proteins including the Class 1 MHC molecules.

Babes-ernst Bodies Metachromiatic granules, or volition; granules of long chains of inorganic phosphate; characteristic of orynebacterium diphtheriae.

Babes-Ernst Granules Metachromatic intercellular deposits of polyphosphtes found in corynebacterium diphtheriae when the bacteria are grown on suboptimal media, Stainreddish with methylene blue or toluidine blue.

Bacillus Calmatte-Guerin (BCG) A attenuated mycobacterium derived from Mycobacterium tuberculosis. The bacterium is used in vaccination against utberculosis. Extracts of the bacterium have remarkable powers in stimulation of lymphocytes and leucocytes.

Bacilius Describing the appearance of rod-shaped bacteria. Bacilli may occur singly (e.g. Pseudomonas) in pairs, or in chains, (e.g Lactobacillus). Some are motile.

Back Mutation A mutation that causes a mutant gene to regain its wild type function (c, forward mutation).

Backbone The atoms in the polymer that are common to all its molecules (e.g. the sugars and phosphates in RNA).

Backcross Method of Breeding A system of breeding carried out by several generations of backcrossing and subsequent selection, the character of the recurrent parent are retained for the most part, and characters from the non-recurrent parent are added.

Backcross Ratio The proportion of heterozygotes to bottom recessive expected in a backcross is given by calculating 2° where n is the number of factors involved, e.g. with two factors 2° = 4 and the expected ratio is 3:1 with 3 factors 2° = 8 and expected ratio is 7:1, etc.

Backflow Preventer A manufactured piping device of the type that has two spring-loaded vertical check valves and one spring-loaded, diaphragm-activated differential pressure relief valve. It is installed in a water supply.

Background Adaptation The agreement between the colour and appearance of the organism and general appearance of an organism and general appearance of the environment in which it lives.

Bacteria A large and diverse group of organism which in terms of number and variety of habits, includes the most successful life form. In nature bacteria are important in the nitrogen and carbon cycles, and some are useful to man in various industrial processes, especially in the food industry. However there are many harmful parastic bacteria that causes diseases such as botulism and tetanus.

Bacterial Community A mixed culture of bacteria including populations of numerous different species.

Bacterial Population A group of bacteria belonging to the same species and exchanging same genetic material among each other, but having few contacts with other groups of the same species.

Bacterial Population A group of bacteria belonging to the same species and exchanging same genetic material among each other, but having few contacts with other groups of the same species.
Bacteriochlorphyll A type of chlorophyll unique to photosynthetic bacteria. There are four forms, designated bacteriophylls a-d All are structurally similar to chlorophyll of plants. The bacteriochlorophylls absorb light at longer wavelengths than chlorophyll a enabling far-red and infrared light to be used in photosynthesis see Photosyntehetic Pigments.

Bacteriocin Any of the proteinaceous toxins liberated by any bacterial strains. These compounds are often named on the basis of organism which produce them, for example Colicins are produced by E. coli and staphylococcus species. Bacteriocins are active against closely related strains, while the producing cells remain immune. These proteins are enclosed by plasmid genes rather than in the main bacterial genome.

Bacteriophage A bacterial virus Virulent phages induce death in a bacterium. Temperate phages live as prophages in a symbiotic association with bacterium, see Temperate Phage and Prophage.

Baceriostat An agent which stops the growth and multiplication of bacteria, but does not necessarily kill them. Usually growth resumes when the bacteriostat is removed.

Bacteriod Irregular shape of bacterium (Rhizobium) associated with symbiotic nitrogen fixation in roots of plant.

Bacticinerator An electrical appliance, capable of generating elevated temperatures, into which metal instruments are inserted and held while sterelisation occurs. Bacticinerators are generally used to facilitate aseptic operations within laminar air flow cabinates, and may replace the Bunsen burner, especially in areas without access to natural gas.

Baculovirus A group of virus tht only infect arthropodas. They are classified according to whether the virions become embedded (occulded) in inclusion bodies. The nuclear polyhedrosis viruses, characterized by polyhedral inclusion bodies, infect Lepidoptera, whereas those that do not form inclusion bodies (nonoccluded) infect a wide range of arthropods. The selectively for their hosts, the ease of recognition and the fact that these viruses do not infect vertebrates make them of interest as potential biological pest agents.

Bakery Foods The main baked product that uses microorganisms in its product is bread. Most breads are made with wheat flour using bakers yeastas the leaving agent. They east gives off carbon dioxide produced during the metabolism of sugars present in the dug. Resulting into an aerated spounge like structure which is fixed during the baking process. They east also contribute to the flavour and the maturation of the dough, particularly through the formation of alcohol which influences the colloidal nature of the flour proteins.

Balance in Drosophila Sex Determination Sex in Drosophila is determined by ratio of X chromosome at autosomes (chromosomes 2 and 3) X?A ratio of 1.0 is female. 0. 5 male, while 0.67 and o.75 are intersexes.

Balance of Nature The natural tendency of plants and animals population resulting from natural regulative proceses in an undistributed environment to neither declinein numbers to extinction nor increase to infinite density.

Balance Polymorphism Retention of both extreme phenotypes in a population, usually resulting when the heterozygote is superior to both homozygotes.

Balanced Lethal (1) A phenomenon caused by the heterozygosity for two closely linked genes (L, and L_2) that are in trans-position. Only the heerozygotes is viable, the two corresponding homozygotes will succumb due to homozygosity L_1 and L_2, respectively (2) Balanced lethality also occurs in cases of complex-heteroxygosity in Oenothera, In such cases each complex is lethal when homozygous: only the heterozygote is viable.

Balanced Polymorphism Two or more distinct types of individuals coexisting in the same breading population, actively maintained by selection.

Balanced Salt Solution A solution containing a mixture of metabolically important inorganic salts in proportions designed in maintain the correct osmotic pressure.

Balbalani Type Chromosome Giant chromosome whether occurring in salivary glands or elsewhere cf, Salivary Gland Chromosomes.

Banding Patterns Chromosomes stained with certain dyes, commonly quinacrine or Giemsa, show a pattern of transverse bands of lights and staining, which is the same in most tissues, is not understood: each band generally represent 5-10% of the length, abôut 107 base pairs, although this is not true for polytene chromosomes which show more than 4000 bands per Drosophilagenome.

Base Analogue (1) A substance differing in minor aspect from one of the regular nucleotide bases. Base analogues may be incorporated into the genetic constitution and as a result induces mutations (2) Purines and pyrimidines that differ slightly in structure from the normal nitrogenous bases, some analogs (e.g., 5-bromouracil) may be incorporated into nucleic acids in place of the normal constituent.

Base Pairing The linking together of the two helical strands of DNA by bonds between complementary bases, adenine pairing with thymine and guanine pairing with cytosine. The specific nature of bases pairing enables accurate replication of the chronomosomes and thus maintains the constant composition of the genetic material. Impairing between DNA and RNA the uracil of RNA pairs with adenine.

Base Ratio The requirement that adenine must always form a base pair with thymine (7) to guanine (G) plus cytosine (C). In DNA the amount of A is equal to the amount of T, and the amount of G equals the amount of C, but the amount of A+T does not equal the amount of C+G. The A+T:G+C ratio is constant within a species but varies between species.

Base specific Ribonuclease An enzyme that cleaves a molecule of RNA at a specific bases; for example RNase T1 is specific for G residues. RNase U2 is A-specific, RNase PhyM cleaves at U and C residues. These ribonucleases are used in RNA sequence analysis.

Basement Membrane A thin layer of jelly or cement that is found at the base of an epithelium

Basidiomycetes A class of true fungi in which the sexual spores basidiospores are borne externally on the end of specialized cells termed basidia.

Basidiospores Spores produced externally by meiosis from adiploid cell called the basidium. Found excusively in the class Basidiomyetes.

Basophil A white blood cell (leucocyte) containing granules that stain with basic dyes. It has a lobed ucleus. Basophils comprises only 0. 5% of all leucocytes. They move aboutin an amoeboid fashion and may ingest bacteria. They are thought to produce heparin ad histamine and are considered to be wandering mast cells, similar to those found in the lining of blood vessels, see also Leucocyte, mast Cell.

Batch Culture It is a discontinuous sytem used for culturing microorganisms, waste treatment or manufacturing products of value, characterized by a single loading or inoculation into fresh medium at the start of the process and a single harvesting stage at the end, usually when the substrate has been exhausted. Organim grown in batch systems generally show a sigmoid growth curve, undergoing exponential growth for a few generation following an initiallag phase. Growth ten ceases, and the cells may lyze die or stop growing.

Batch Recycle Culture A discontinuous culture technique in which the cells are harvested at the end of the growth phase and reintroduced into fresh medium, which result into a new cycle of growth being initiated. This process is used in the manufacture of secondary products in order to increase the substrate use efficiency since raw materials are not used to generate new cells for each s in the case of a conventional batch process.

Batesian Mimicry The resemblance of an innocuous species to another species which is protected from predatory types by unpalatability or by other qualities.

Bathmic force A hypothetical, inherent, adaptive growth force supposed to bring about variation and to control evolution.

Batrachotoxin Refers to a neurotoxin from the skin of the columbian arrow poison about a selective and irreversible increase in the permeability of nerve membrances to sodium ions.

Bayer's Junction Sites of adhesion between the outer and cytoplasmic membrances of gram negative bacteria: each cell has 200-4000.

Bed Volume The total volume occupied by the chromatographic packing material. In column chromatographic packing material. In column chromatograpy, it can be determined by filling the tube to the original level of the bed with water an measuring the volume.

Bee Toxin A defenive secretion which is produced in an abdominal gland of queen and worker bees and delivered by the sting. It contains three types of active principle; (1) biogenic amines, including hitamine, which ceauses pain and dilae blood vessles, allowing wider penetration; (2) biologically active peptide such as mellitin apamin, and (3) enzymes like hyaluronidase ad phospholipase A B. t. is applied interactously or percutaneously for treatment of neuralgias, rheumatism and allergies.

Belt Press A mechanical device used to dewater sludge, plup derived from the extraction of the plant material etc. The press consists of one or two endless belts which are rotated continuously under tension. The slurry is passed between the belts, or between the belt and a perforated drum, resulting in moisture squeezed out leaving a semisolid cake.

Benedict's Solution A mixture of sodium citrate, sodium carbonate and copper sulphate used for testing for reducing sugars in solution, a red precipitate of copper oxide is formed when benedict's solution is boiled with a solution containing reducing sugars. Benedit's solution thus reacts similarly to Fehlin's solution, but is more sensitive to low concentration of sugars, with which it gives green to yellow precipitate, it is used to test for the presence of sugar in urine in screening for diabetes.

Benedict's Test A biochemical test for reducing sugar based on an alkaline copper sulphate solution and sodium citrate. Reducing sugars produced a coloured precipitate which varies from green through yellow to red with increasing amounts of sugar.

Benzalkonium Chloride or Zephiran Chloride An alkyldimethyl-benzylamines employed as a disinfecting agent for plant material.

Berk-sharp Technique A technique of genetic mapping in which messenger RNA is hybridized with single-stranded DNA and the non-hybridised with the messenger is then charracterised by electrophoresis. Also known as S1-mapping.

Beta Pleated Sheet A type of protein structure in which polypeptide chains run close to each other and are held together by hydrogen bonds at right angles to the maintain. The structure is folded in regular 'pleats' fibres having this type of structures are usually composed of amino acids with short side chains. The chains many run in the same direction (parallel) or opposite directions (antiparallel).

Beta-counter A device for quantifying the radioactive decay from a beta-emitting radioisotope. This can br either a Geiger-Muller counter or a liquid scinetifllation counter.

Beta-glalactosidase An enzyme which will cleave lactose into glucose and glalctose. The most commonly used galactosidase gene is from the Escherichia coli lac operon.

Bial's Reagent A mixture of orcinol in concentrated hydrochloric acid with 10% iron chloride added, used in testing for pentoses. A positive reaction is shown by green coloration produced when the test material is boiled with the reagent.

Bias A consistent departure of an observed quantity from its proper value owing to error of observation or estimation.

Bifunctional Vector or Plasmid A DNA molecule able to replicate in two different organisms e.g. in E. coil and yeast or E. coil and Streptomyces, these molecule are thus able to "shuttle" between the two alternative hosts and are therefore also known as shuttle vectors. It is usual to 'grow' DNA in E. coil for genetic transformation of the alternative host.

Bilateral Symmetry The condition in which only the right and left sides, divided exactly down the back, are mirror images of each other.

Binal Refers to becteriophages with a combination of two different symmetryes: for example, bactriophages T, with a cuboidal head and a helicaltail.

Binding Proteins Soluble proteins which specifically and reversibly bind a variety of substances (ncluding sugars, amino acids, inorganic ions and vitamins) but which appear to have no enzymix activity. Binding proteins are believed to occur in the periplasmic space of many gram-negative bacteria, and have also been found in certain yeast. They are believed to play some role in the transport of the substances bound, but the precise nature of this role remains uncertain, it is thought that they may act as sites of the recognition for the substance to be transported. Osmotic shock has been used to release the binding proteins from Gramnegative bacteria.

Binding Site A specific arrangement of atoms or molecules that is recognized by and forms the point of attachment for, an ion compound, antibody, virus cell or organism on another structure or organism.

Binomial Experiment An experiment with the following properties: (1) Experiment repeated under identical conditions (trials): (2) Dichotomous outcomes (success and Failure): (Prob (success] = p and is constant from trial to trial, and Prob[Failure] = 1-p = q. (3) Trials are independent.

Binomial Consisting of two names; for example, the binomal nomenclature of biology which gives the name of the genus followed by the name of the species.

Bio-gelp A cross-linked polyacylamide support material used in gel filtration chromaogrpahy. The gel is prepared by copolymerization of acrylamide and N.N'-methylendiacrylamide in aqueous solution. It is commercially available in ten porosities up to a molecular exclusion limit o 4 * 105 for globular proteins and 1.5 * 103 for linear polymers.

Bioautography A method for detecting small amounts of one or more substances (e.g. vitamins) in a complex mixture such substances being essential requirements for the test organism (s) used. The components of the mixture are initially separated by chromatography. The presence of a particular component (in a given chromatographic band) is than detected by extracting the band, with an appropriate solvent, and adding the extract to a medium which lack the relevant growth factor. The medium is inoculated with an appropriate test organism, and incubated. Growth indicates hat the particular component way present in the original mixture.

Biocatalyst A catalyst that cosists of, or is derived from, a living organism or tissue, or cell culture, Biocatalyst may be categorized as follows: 1. Cells (a) Growing The use of growing cells in suspension as biocatalysts is generally termed fermentation. Such processes may be distinguished as batch, fed batch or continuous (b) Non-Growing systems include the use of suspensions, cells retained within a semipermiable membrane or cells immobilized in or on a solid support. Such immobilized cells maybe used in a fixed bed, expanded bed, fluid sized bed or fully mixed system. 2 Enzymes (a) Soluble Enzyme System, Soluble enzymes of varying purity may be used in solution, generally in a batch system and are lost at the end of the reaction (b) immobilized Enzymes. Enzymes may be entrapped, adsorbed on to surfaces or cross-linked to support materials and then used in fixed bed, expanded bed or fully mixed systems.

Biochemical Genetics The branch of genetics concerned with the inheritance of genetic difference in the ability or inability to synthesize or metabolize certain chemicals.

Biochemical Mutants Usually bacterial or fungal mutants which cannot grow on minimal media (containing sugar, mineral nutrients an biotin) unless supplemented by the specific amino macid or vitamin which it is unable to synthesise, they are also called auxotrophs and were first discovered by Beadle in Neurospora crassa, such biochemical mutants have been also reported in some angisperms i.e. Arabidopsis thaliana (cruciferae).

Biochemical Oxygen Demand (BOD) The standard measure for determining the level of organic pollution in a sample of water. It is the amount of oxygen used by microorganisms feeding ion the organic material over given period of time. Sewage effluent must be diluted to comply it can be disposed of into clean rivers.

Biochemical Taxonomy The use of chemical characteristics to help classify organisms; for example, the Asteroideae and Cichoriodeae, which are the two main divisions of the plant family compositae, are separated by the presence or absence of latex, this area of taxonomy has increased in importance with the development of chromatograpy, electrophoresis, serology and other analytical techniques.

Bioconversion The transformation of matter, from one compounds occurring in living organisms.

Biodegradation A process in which material is broken down to its chemical constituents by the action of living organisms. The term is often of living organism. The term is often restricted to the breakdown of wastes released into the environment in general biodegradation is regarded as a desirable process in contrast to Biodeterioration.

Biodyne A form of activated nylon filter which can be used in the place of nitrocellulose in the Southern blotting procedure. Membrane is that they are more robust than nitrocellulose and the same filter may be used in a laıge number of consecutive hybridizations.

Bioengineering (1) The design, manufacture and use of replacements of aids for body parts or organs that have been removed or are defective, e.g. artificial limbs, hearing aids, etc. (2) The design manufacture and use of equipment for industrial biological processes, such as fermentation.

Biofilter An item of equipment used in the treatment of sewage and the purification of industrial waste water that use active biomass grown on a solid support. Biofilters are circular or rectangular tanks made from a concrete containing a grade biosupporut medium, usually clinker, slag stone or gravel, packed to a depth of 1.8 m. they include septic tanks, low-rate percolating filters, double filtration systems and high rate filtration systems.

Biofuel Cell A device that use biochemical reaction in association with the principles of a fuel cell. Biofuel cells may be classified as direct or indirect fuel cells, in an indirect cell the electroactive species, such as hydrogen is generated by anaerobicdigestion or photolysisof water and subsequently oxidized at the anode of the conventional hydrogen/oxygen fuel cell. In a direct cell, a redox protein isused as an intermediate in the direct electron transfer from a substrate to an electrode. Forinstrance, the complete oxidation of methanol, catalyzed by methanol dehydrogenase and format dehydrogenase, can be used to generate a current using a complex organic compound such as N, N, N', N'-tetramethyl-p-phenylaminediamine as a mediator. Such systems are still at the experiment stage, but have been demonstrated to give a steady current for over 24 hours. Biofuel cell could use the organic matter in food processing and other industrial waste streams as a sustrate, thus combining water treatment with energy production.

Biofuel A solid, liquid, or gaseous fuel obtained from biological raw materials biomass). The conversion process may be accomplished by thrmochemical or biological means, he major biofuels produced biologically are biogas generated by anaerobic digestion (Biomenthanation) and fuel ethanol generated by a yeast-based fermentation of molasses, sugar cane juice or hydrolysed starch.

Biogas Plant An anaerobic digester together with ancillary equipment. These additional pieces of equipment include systems for storing and processing the influent (substrate), storing and/or treating the biogas produced, using the gas to produce heat or power, and storing and/or treating the effluent.

Biogas A mixture of methane and carbon dioxide along with traces of other gases such as hydrogen, nitrogen, hydrogen sulphide and water vapour, which is produced during anaerobic digestion or biomethanation.

Biogenesis The theory that living things originate only from other living things as opposed to nonliving matter, the theory became accepted as a result of the work of Redi and Pasteur, ho showed that dirt, for example, does not itself produce bacteria or maggots only come from spores or eggs already existing in the dirt. This theory satisfactorily explains the occurrence of existing in the dirt. This theory satisfactorily explains the occurrence of existing organisms, but not the origin of the first organisms.

Biogenetic Law The hypothesis that ontogeny recapitulates phylogeny.

Biogeochemical Cycles Movement of elements through living organisms and the physical environment.

Biogeochemistry The science that deals with the role of living organism in affecting the chemistry of the earth and its atmosphere.

Biogeography The scientific study of the geographic distribution of organisms. Ecological biogeography is concerned with the habitats in which organisms live, historical biogeography with the complete geographic ranges of organisms and the historical circumstances that determine the ranges.

Bioinformatics in India Nine Distributed Information Centres (DICs) have been established in India along with 14 user centres. The nine DCIs located at the following institutions: (1) Indian Institute of Science (IISc), Bangalore, (2) Madurai Kamraj University (MKU), Maduri, (3) Bose Institute, Calcutta, (4) Jawahar Lal Nehru University, New Delhi, (5) Poona University, Poona, 6) Centre for Cellular and Molecular Biology (CCMB), Hyderabad, (7) National Institute of Immunology, New Delhi, (8) Institute of Microbiological Technology, Chandigarh and (9) Indian Agricultural research Institute, New Delhi. These DICs are linked wit each other and wit the Central Information Network at DBT in New Delhi which access to International Databases in biotechnology.

Biologic Oxidation Any chemical reaction occurring within a cell that results in a release of energy, respiration. Oxidation results from the addition of oxygen, or the loss of hydrogen and electrons.

Biological Clock Mechanisms that allow expression of certain biological structures (genes) at periodic intervals. May involve alpha and anti-alpha factors.

Biological Controls (1) The use of natural predators, instead of chemicals, to control pests, The most famous successful example was the introduction of the grey moth, cactobalastic cactorum, into Australia to control the prickly pear, Optimum internis which was overrunning vast tracks of land. (2) Biological pest suppression in its narrow, classical sense, usually restricted to the introduction, by man, or parasitoids, predators and/or pathogenic microorganisms to suppress population of plants or animals pests; See Biological Insect Pest <Suppression<Natural Control.

Biological Oxygen Demand (BOD) A measurement of the amount of oxygen required for the microbial decomposition of the organic matter present in water.

Biological Race (1) Such a race may be said to exist wnere the individuals of a species can be divided into groups usually isolated to some extend by food preference occurring in the same locality and showing definite defferences in biology, but with corresponding structural differences either few or inconstant or completely absent (2) In disease the term denotes a strain which though morphologically indistinguishable yet differs from the normal in its pathogenicity.

Bioluminescence The production of light by living organism. Bioluminescence is found in many marine organisms especially deep-sea organism especially deep-sea organisms. It is also a property of some insects, e.g., fireflies and certain bacteria. The light is produced as result of a chemical reaction whereby the compound luciferin is oxidized. An enzyme, inciferase, catalyses the reaction in which ATP supplies the energy. There are several varieties of luciferin.

Biomass Concentration The weight of cells, expressed on a wet weight or dry weight basis, per unit volume of medium or fermentation volume.

Biomass Hold Up The living microbial or cell culture mass present in a fermenter or bioreacter, or associated with support particles.

Biomembrane A structure having lipids, glycolipids, proteins and glycoproteins, bounding the cell (cell membrane) or subdividing it into compartments. It is sheet-like structure about 60 to 100 angstrome thick, the lipids have hydrophilic "head" groups and hydrophobic "tail" regions.

Biomethanation The anaerobic fermentation of organic material to produce biogas (a mixture of methane and carbon dioxide).

Biopesticide A pesticide in which the active ingrediant is a virus, fungus or bacterium, or a natural product derived from a plant source.

Biophore Weismann postulated that each determinant (gene) is made up of a complex of biophores which he regarded as the smallest ultimate units of living matter. A biopere is the equivalent of the biogen of Verworn.

Biophysics The study of the physics of biological systems. It can be subdivided into three main areas; molecules and cellular biophysics, which is concerned with membrane properties, bioelectric phenomena and the confirmation and conformational changes in biopolymers; radiation biophysics, which is the study of the changes in biopolymers; radiation biophysics, which biophysics is the study of the changes in cells and macromolecules caused by radiation; and therapy and medical technology.

Bioplasm Protoplasm; living cell substance including the nucleus and cytoplasm.

Bioplast A hypothetical, self-propagation vital unit of granular structure in the cytoplasm.

Biopolymer, biomacromolecule A naturally occurring substance which can be regarded as a chain (polymer) of identical or similar subunits Synthesis of biopolymers begins with the transfer of an activated monomer to a primer molecule (initiation). It continues with the addition of more activated monomers, one at a time (elongation) At some point the growth terminates for reasons which have not been entirely elucidated. The Biopolymer is not necessarily complete at this point (Posttranslational modification); it may be further modified.

Biopolymer Any of the different types of large molecules formed by living organism including nucleic acids, proteins, polysaccharides ad lipids.

Bioreactor A vessel used to carry out a biological reaction. The term may be applied to the reactor used in the culture of aerobic cells, or to columns or packed beds of immobilized ell or enzymes. Bioreactors containing immobilized biocatalyst to carry out biological transformation such as hydrolytic reactions, isomerisation decolourization etc. In some forms the bioreactor may be used as biosensors. In such applications, a stream of samples, buffer and reactants move through the immobilized bed. Sensors are placed before and after the after the reactor and the difference determined.

Biosensor A device that uses an agent of biological origin, or a biological principle, for detecting or measuring a chemical compound.

In general biosensors use isolated enzymes, immunosystems, tissues, organelles or whole cells as catalysts.

Biosystematics The area of systematic in which experimental taxanomic techniques are applied to investigate the relationship between taxa, such technique include sereological methods, crossing experiments and cytological examination in addition to the more established procedures of comparative anatomy, evidence from ecological studies may also be brought to bear.

Biotechnology Biotechnology is defined as the applications of scientific and engineering principles to the processing of material by biological agents to provide goods and services. ***British Biotechnologist***: Application of biological organisms, system or processes to manufacturing and service industries. ***The Spinks Report*** defined biotechnology as the application of biological organisms, systems or processes to the manufacturing and service industries. ***Japanese Biotechnologists***: 'A technology using biological phenomena for copying and manufacturing various kinds of useful substances.

Biotin [$C_{10}H_{16}O_3N_2S$ mw 244.31] Part of the vitamin B complex. It is also called vitamin H. It is present in all living cells, bound to polypeptides or proteins, and is of import in fat, protein and carbohydrate metabolism. It is a common addition to plant tissue culture media.

Biotinylated –DNA ADNA molecule labeled with biotin by incorporation of biotinylated-dUTP into a DNA molecule. It is used as a non-radioactive probe inhybridisation experiments such as southern transfer, the detection of any hybrids uses a complex of streptavidin-biotic-horseradish which will give a flurescet green colour when hybrids have formed.

Biotinylated-d UTP A nucleoside triphosphate with the vitamin biotin attached, via a specer arm, to the d UTP, It can incorporated into a DNA molecule by nick translation and then the biotnylated-DNA formed from such a reaction can be used as a probe in a hybridization experiment. It is a nonradioactive alternative to labolling with E E Ep.

Biopolar Cells A class of retinal interneurones, named after their morpholoy, which receive in put from the photorecpuis and send it to the ganglion cells. Bipolar cells are non-spiking; their response to light is evenly graded and shows lateral inhibition.

Birboim-doly Procedure A rapid method for the purification of plasmid DNA, often used to screen recombinant colonies to determine the size of a DNA fragment inserted into a vector. The technique uses an alkaline denaturation step followed by a raid renaturation to achieve the removal of the majority of the chromosomal DNA and most of cells RNA. It can be used for minipreps or scaled up to give large quantities of fairly pure plasmid DNA. The produced at the end of the procedure is usually sure enough to be digested by a restricting enzyme.

Bisexual Bearing both the male and female sexes.

Blade The lamina or expanded part of a leaf or petal. (2) The cutting end of knife or scalpel.

Blending Inheritance Inheritance in which the characters of the intermediate level in the hybrid with no apparent segregation in later generations.

Blocking Antibodies (Incomplete antibodies; non-agglutination antibodies) antibodies which fail to agglutinate homologous antigens even though antigen-antibody combination may occur Lack of agglutination may be due e.g. to an insufficiency of effective combining with the antigen, blocking antibodies prevent ('block') agglutination by homologous agglutinating antibodies. The combination of antigens with blocking antibodies may be detected e.g. by nan Antiglobulin consumption test.

Blood culture (Bacterial) A procedure used for detecting and identifying bacteria in blood. Pathogenic bacteria commonly appear in the blood during the course of certain diseases, e.g. the early phase of Typhoid fever Since blood-borne bacteria may be present only in very small numbers, attempts are not generally made to culture them by direct plating methods.

Blood Group Antigen Refers to the specific oligosaccharide structures attached to glycoprotein in the membranes of blood cells which are recognized as antigens by the immune systems of other individuals or organisms. The antigens get attached to the protein glycophorin in erythrocytes and to both proteins and lipids in other parts of the body. In humans, five systems of antigems could be identified, the ABO systems. Only the ABO and rhesus systems affect blood transfusion between humans; the other systems have been identified using animal antibodies against human blood.

Blood Groups The classification of types of blood based initially on the occurrence of clumping (agglutination) of the red blood cells when blood of incompatible groups is mixed.

Blood Pasma The straw-coloured liquid that remains when all the cells are reoved from blood. It consists 91% of water and 7% of proteins, which are albumins, globulins (mainly antibodies), and prothrombin and fibrinogen (concerned with clotting). Plasma also contains the ions of dissolved salts, especially sodium, potassium., chloride, bicarbonate, suplhate, and phosphates. Plasma is slightly alkaline (pH 7.3) and the protein and bicarbonat act as buffers to keep this constant. It transport dissolved food (amino acid fat and fatty acids) excretory products, dissolved gases hormones and vitamin. Most of the body's physiological activities are concerned with maintaining the correct concentration and pH of all these solutes, since plasma supplies the extra cellular fluid that is the environment of all the cells.

Blood Vessel A tubular structure that is the part of the blood vascular system Blood vessels include .veins, arteries, venules, arteroids, capillaries and sinusoids.

Blood-Brain Barrier The blood vessels of the brain (and the retina) Are much more impermeable to large molecules (like antibodies) than blood vessels elsewhere in the body. This has important implications for the ability of the organism to mount an immune response in these tissues, although the basis for the difference in endothelial permeability is not well understood.

Blood The fluid circulated by the heat through the circulatory system. Carries oxygen, nutrients, hormones, carbon dioxide, wastes, and other substances.

Booster Dose Additional injection of an antigen to maintain antibody production at its peak.

Bottle Cells The first cells to migrate in waeds at the blastopore during amphibilian gastrulation. The "neck" of the bottle is at the outer surface of the embryo.

Breaking First Hypothesis The theory that chromosomes can suffer fragmentation after which the fragments may either remain loose, or become reattached to restore the original gene arrangement, or may unite wit other fragments to from new aggregates.

Breakage and Reunion The model of crossing over by physical breakage and crossway reunion of completed chromatide during meiosis. This model have been recently shown to be applicable in atleast one cases on the molecular level-crossing over between phage-DNA molecle proceeds by breàkage and reunion.

Breakage-reunion A model of recombination in which the homolegus DNA molecule breaks at identical sites and reanneal with each other.

Broth Sugars ('sugars') Bacteriological media used to determine the range of carbohydrate metadolized by a given strain of bacterium. Except in that they are broth-based, broth sugars are similar to peptone water sugars both in composition and mode of usage.

Brownian Movement Random movement of bacterial cells and small particles due to bombardment by water molecules.

Brucella Species Small, gram-negative, coccoid to rod-shaped cells; causative agents of brucellosis in animals and humans.

Bud Primordium In plants, a small mass of potentially meristematic tissue found in the angle between the leaf stalk and the shoot apex Will give rise to a lateral branch under appropriate conditions.

Bud-mutation: Bud Sport A mutation in the bud. Such a mutated bud may grow into a branch or flower differing in one.or more characters from the parent plant.

Bud-Sport A somatic mutation or variation arising in bud and producing an abnormal branch.

Budding An asexual reproductive process in which the parent cell retains its identity while the offspring (bud) is produced with a newly synthesizedcell wall attached to parent cell; also a process sin which an enveloped virus is released through the plasma membrane of a vertebrate cell.

Buffer A substance in solution, or system that resists pH change (despite addition of some small amount of acid or base) and withstand shock; as buffered media can resist pH drift. The degree to which pH change is resisted is a measure of the solution buffering capacity.

Buffy Coat Thin yellow white layer of leucocytes on top of the mass of red cells when whole blood is centrifuged.

Bufotizin The main toxin is the venom of the European toad Bufo vulgaries. It is steroid derivative.

Burst Size The number of phage particles released from a bacterium after maturation of the phage. The burst size is usually in the range of 50 to 200 new bacterio phages.

Butanol An alcohol, produced in some bacterial fermentations used as an industrial solvent.

Buttermilk A fermented product obtained by culture of Steptococcus lactis, S cremoris and/or Leuconostoccremoris in pasteurized skimmed milk.

Butyeric Acid An organic acid produced in some bacterial fermentations.

Butyric Acid Fermantation A type of fermentation, carried out by certain saccharolytic clostridium species (e.g., C., butyricum) in which glucose is fermented with the formation of acetic acid, butyric acid, $CO_2$and H_2.

C

C_3 Photosynthesis The form of photosynthesis in which 3-phosphoglycerate is the first stable product, and ribulose bisphosphate is the CO_2 receptor.

C_4 Photosynthesis The form of photosynthesisin which oxaloacetate is the first stable product, and phosphoenolpyruvateis the CO_2 acceptor, C_4 plants also perform the reactions of C_3 photosynthesis.

Calcium Hypochlorite [$Ca[OCl]_2.4H_2O$ mw 215]. An algicide bactericide and fungicide capable of disinfecting and bleaching. It is used in dilute solution [5-10% w/v] as a plant tissue disinfectant often when agitating or using a vacuum. Tissue damage occur if contact is prolonged. Through washing with sterile water generally follows treatment.

Calcium Pantothenate or Pantothenic Acid [$Ca[C_9H_{16}NO_5]_2$ mw 476.54]. The calcium salt of vitamin B_5. An occasional vitamin additive to plant tissue culture media. Solutions of this vitamin are not stable to autoclaving.

Callus Culture The cultivation of callus, usually on solidified medium and initiated by inoculation of small explants or sections from established organ or other cultures (inocula). Callus may be maintained indefinitely by regular subdivision and subculture. It may be used as the basis for organogenetic [shoot, root] cultures, cell cultures or proliferation of embryoids.

Calorie The amount of heat required to raise the temperature of one gram of water by one degree Celsius (1 °C) from 14.5 °C to 15.5 °C. In butrition studies, "calorie" refers to the kilocalorie (1 kcal=1000 cal). The amount of heat required to rise the temperature of one kilogram of water by 1 °C.

Calvin Cycle The pathway which reduces CO_2 to carbohydrate (cH_2O) in the stroma of chloroplasts an the cytosol of autotrophic bacteria. Eukaryotes and blue-greens use $NADPH_2$ as a hydrogen donor; other auto-trophs use $NADH_2$. Some sulphur bacteria supplement the Calvinçycle by forcing the Krebs cycle to run backwards using reduced ferredoxin.

Camp (cyclic AMP) A compound, formed from ATP, that mediates the effects of numerous animal hormones. Also needed for the transcription of catabolite-repressible operons in bacteria. Used for communication by cellular slime molds.

Candida Albicans Yeast-like fungus; a member of the normal flora of humans, and associated with certain infections, e.g., thrush.

Capping A process by which cell surface molecules are caused to aggregate (usually using antibody) on the cell membrane.

Capsule A mucilaginous envelope surrounding certain microorganisms, external to the cell wall; frequently consists of polysaccharide.

Carbon Budget The amount of atmospheric carbon (from carbon dioxide) incorporated into organic molecules by plant.

Carbon Source A source of the non-metallic element carbon (C); as organic substances like sugars taken up and metabolized by plant tissue cultures.

Carcinogen An agent (chemical, physical, or biological) that an cause malignant tumor formation.

Carnivore An organism that feeds on animal tissue. (Contract with detritivore, herbivore, omnivore.)

Carotenoid A yellow, orange, or red lipid pigment commonly found as an accessory pigment in photosynthesis; also found in fungi.

Carrier A person who harbors and spreads disease-producing microorganisms without having the over symptoms of the diseases.

Carrier A person who harbors and spreads disease-producing microorganisms without having the overt symptoms of the disease.

Carrying Capacity In ecology, the largest number of organisms of a particular species that can be maintained indefinitely in a given part of the environment.

Casamino Acid An amino acid obtained through the digestion [acid drolysis] of the milk protein casein. Protein digest yields casamino acids and other substances. These are undefined constituents of some plant tissue culture media.

Cell Count The number of cells per unit suspension volume or callus weight. Tissue is treated with chromic acid [5-8%] or pectinase [0.25%] for up to 15 minutes followed by mechanical dispersion, then cell numbers are estimated with a hemocytometer [haemocytometer].

Cell Culture Term used to denote the maintenance or cultivation of cells in vitro including the culture of single cells. In culture, the cells are no longer organized into tissues.

Caseation The process by which a tubercular lesion develops into a coagulated cheesy mass.

Casein The principle protein (phosphoprotein)of milk; converted to calcium caseinate by rennin, leaving the clear residual fluid, whey.

Cassava Any of several plant of the spurge family grown in the tropics for their fleshy edible rootstocks, which yield nutritious starch.

Catabolism Negative metabolic balance; destructive metabolism; the breakdown of complex compounds, with the liberation of energy.

Catalase An enzyme which catalyzes the breakdown of hydrogen perioxide to yield water and oxygen.

Caudal Pertaining to the tail, or to the posterior part of the body.

Cell Count The number of cells per unit suspension volume or callus weight. Tissue is Treated with chromic acid [5-8%] or pectinase [0.25%] for up to 15 minutes followed by mechanical dispersion, then cell numbers are estimated with a hemocytometer [haemocytometer].

Cell Cycle The process of cell division which is divisible into four phase; C1, S, G2 M. DNA replicates during the S phase and the cell divides in the M (mitotic) phase.

Cell Hybridization Formation of synkaryons, viable cell hybrids produced through cell fusion they are identifiable by their increase chromsome number compared to the parent cells and the possession of characterise found in one or another of the parental cells.

Cell Selection Selection within a group of genetically different cells; involves competition between cells often under some stress. Selection criteria may involve cell viability; a unique phenotype or biochemical activity; or another basis for choice. Select cells or cell lines are generally relocated to fresh medium for continue selection and often are exposed to an increased level of the stress agent. The final objective is usually to regenerate plants from those select cells, in hope that the plants will exhibit the traits selected for at the cellular level.

Cell Strain Cells having specific properties or markers derived either from a primary culture or cell line are referred to as a cell strain; a characterized cell line derived by selection or cloning.

Cell Theory The theory, well established, that organisms consist of cell, and that all cell come from preexisting cells.

Cell Transformation A permanent alteration of the cell phenotype presumed to occur via an irreversible genetic change; may be spontaneous or induced by chemical of viral action, or by exposure to radiation. Transformation is often characterized (but not defined) by the emergence of an established cell line from a primary explant, an alteration in typical morphology, loss of contact inhibition, abnormal karyotype, changes in viral susceptibility, changes in antigenic properties, neoplastic properties, and the loss of anchorage dependency.

Cell Wall A relatively rigid structure that encloses cells of plants, fungi, many protests, and most bacteria. The cell wall gives these cells their shape and limits their expansion in hypotonic media.

Cellular Immune System That part of the immune system that is based on the activities of T cells, Directed against cancer cells, parasites, fungi, intercellular viruses, and foreign tissues (grafts). (Contrast with humoral immune system.)

Cellular Immunity Immunity developed by certain cells that have acquired an increased ability to ingest or destroy specific foreign substances.

Cellulitis Rapidly spreading infection, involving the skin and the subcutaneous and lymphatic tissues; can be caused by a β-hemolytic streptococcus.

Central Lymphoid Organs Lymphatic organs where lymphoid cell subpopulations become committed to specific functions; the thymus and bone marrow for T and B cells, respectively; also called primary lymphatic organ.

Centriole A cylindrical organelle in animal and some protist cells, containing microtubules in a characteristic arrangement consisting of nine triplets A centriole is at each pole of the spindle apparatus in dividing nuclei (other than in vascular plants.)

Celebellum The brain region that controls muscular coordination; located at the anterior end of the hindbrain.

Chemically Defined Medium A nutrient solution for culturing cells in which each component is of known chemical structure. Distinct from serum-free where other incompletely characterised constituents may be used to replace serum.

Chemiosmotic Hypothesis According to this hypotheis. ATP formation in mitochondria and chloroplasts results from a pumping of protons across a membrance (against a gradient of electrical charge and of pH), followed by the return of the protons through a protein channel with ATPase activity.

Chemotherapeutant A chemical used to pretreat diseased source plants prior to excision or incorporated into media to support some therapeutic objective; as malachite green or virazole [ribavirin] are used to eliminate virus in meristem tip culture. The process is *chemotherapy*.

Chimera or Chimaera A plant or tissue composed of more than one kind of genetic tissue. In a periclina chimera one genotype occurs in a superficial layer, covering a genetically different core. In a mericlinal chimera one genotype occurs in a localized part of the plant and the rest is occupied by another regenotype. In a sectorial chimera two genotypes share distinct sectors of the plant. Chimeras can occur naturally; be established artificially by grafting; or developed through somatic mutation by chemicals such as colchicines, or by other means in plant tissue cultures.

Chlamydiae Obligate intracellular parasites which appear to be related to gramnegative bacteria, and which have a complex method of reproduct on that sets them apart form Rickettsiae; they are not transmitted by an arthropo vector.

Chromatic Adaptation The ability of some algae to modify the kinds and amounts of different photosynthetic accessory pigments. Producing those that absorb the wavelengths of light maximally available in their environment.

Chromatidia Particles of chromatin in the cytoplasm surrounding the nucleus.

Chromatography Anyone of several techniques for the separation of chemical substances, based on differing relative tendencies of the substances to associate with a mobile phase or a stationary phase.

Chromosomal Break A break in the chromosome that serves as the basis for the repatterning of the physical structure of chromosomes. Inverted blocks of genes inversions and the exchange of chromosomes arms are two commonly observed rearrangements each requires a minimum of two breaks.

Chromosomal Bridge A centric chromatid the two centromeres of which move towards opposite poles at the first division so that the chromatid remaine a 'bridge' between the anaphase chromosomes.

Chromosomal Fibres Fibre bundles connecting the tiny granules called "spindle spherules" (important parts of the centromere) with the poles. These chromosomal fibres have by many authors been regarded as intimately involved in chrmosmal movement as traction fibres for the chromosomes.

Chromosome The term proposed by Waleyer for the individual threads within the cell nucleus. The expression lamp brush chromosome describes thè appearance of the chromosomes at the diplotene stage of meiosis in the oocytes of some amphibia, when the chromatids extend from each chromosomes to form lateral loops.

Chromosome Cycle All the changes which take place in the chromosomes of an organism from fertilization to gametogenesis.

Chromosome Mapping The location by linking test of genes in the different chromosomes in linear fashion, by the strength of the linkage of the genes.

Chromosome Matrix A change in the number or arrangement of genes in a chromosomes. If chromosome segments break away during nuclear division they may rejoin the chromosome the wrong way round, giving an inversion. Alternatively, they may rejon a different part of the same chromosome, or another chromosome, giving a translocation. If the segment becomes a lost, this is termed a deficiency or deletion; it is often fatal. A part of a chromosome may be duplicated and occur either twice on the same chromosome or on two different nonhomologus chromosomes; this is a duplication. Chromosome mutations can occur naturally but their frequency and is increased by the effect of X-ray and chemical mutagens.

Chromosome Theory The theory according to which Mendelian factors or genes are located on chromosomes. The chromosomes. The chromosomes are the physical basis for heredity.

Chromosome Walking A method used to identify which clone in a gene bank contains a desired gene or sequence which cannot be selected for easily. It is the identification of fragments with overlapping sequences may be a key to the reconstruction or characterization of large chromosome regions. This is achieved by the technique popularly described as chromosome walking. The technique of chromosome walking involves the following steps. (1) From the genomic library a alone of interest (identified by a probe is selected (2) One end of the close ins subcoloned. (3) the subcloned fragment of the selected clone may be hybridized with other clones in the library and a second clone hybridzing with the subclone of the first clone is identified due to presence of overlapping region (4) The end of the second is then subcloned and used for hybridization with other clones. This helps in identify in a third clone having overlappin region with the subcloned end of the second clone. (5) Identified third clone is also subcloned and hybridzed with clones in the same manner and the procedure may be continued. (6) Restriction map of each selected clone is prepared and compared to know the region of overlapping, so that identification of few overlapping restriction sites resembles the walking along the chromosome or along a long chromosome segment. Regions of chromosomes approaching 1000 kb have been mapped following the above technique. Restriction maps of entire chromosomes can be prepared in this manner following the techniques of chromosome walking.

Chromosomin A special protein in cell nuclei, thought by some to be the principal constituent of chromosomes (others hold nucleoprotein to be the principle constituent).

Chronic Granulamatus Disease Disease, usually fatal in childhood, in which the production of hydrogen peroxideby phagocytes does not occur because of a lesion in an NAD-or NADPH-dependent oxidase. catalase-negative bacteria are not killed and there is no luminal-enhanced chemiluminescence when the cells are tested. The absence of the oxygen-dependent killing mechnism is not itself fatal but seriously compromises the primary defence system. At least three separate lesions can cause the syndrome. The commonest being a defection plasma membrane cytochrome.

Circular Dichroism differential absorbtion of right-hand circularly polarized fight resulting from circularly asymetry involving a cromophore group. CD is used to study the confrontation of proteins and nucleic acids in solution.

Cis-activation Activation of a gene by an activator located on the same chromosome i.e. not by a diffusible product.

Cis-dominant Ability of a locus to affect other genes on the same DNA molecule in a dominant way.

Cis-trans Effect The phenomenon resulting from recombination within a gene (cistron), in which a mutation is only expressed in the phenotype. If the mutant pseudoalleles are on different homologus chromosomes.

Cistron The term derived form cis and trans, for the functional unit of the hereditary material defined by the phenotype of the trans heteorokaryon or heterozygote for two recessive mutations. If this phenotypeis mutant, the mutatons are said to be alleles and to belong to the same cistron; if the phenotype is normal the mutations are said to be non-allelic and to belong to different cistrone.

Classical Genetics Mendelian genetics a study of genetics on an organismal rather than cellular level.

CMI (cell-mediated immunity) A term used to refer to immune reactions which are mediated by cells rather than antibody or other humoral factors.

Co-ordinate System Used to locate and locate items in relatively large areas, as original single cell inoculum sites. A T-shaped marker on the bottom of the petri dish facilitates precise placement of the dish over labeled grid.

Co-repressor The end product of a biosynthetic pathway that combines with aporepressor (a protein) to control an opern. It is a low molecular weight compound that unites with a protein (the repressor) to prevent transcription in a repressible operon.

Coadaption The harmonious interaction of genes; the selection process by which harmoniously interacting genes become established in a population.

Coagulum A curd or clot formed as result of coagulation. The term is used in the manufacture of cheese to refer to the initial product of milk solidification brought about by microbial or enzyme action.

Coconut Milk or Coconut Water Liquid endosperm from the centre of the coconut seed. A complex undefined addendum of variable quality and effects in some nutrient solutions [2-15%, v/v] for plant tissue culture. It has growth promoting effects and cell division factors. It is replaceable in some cases by cytokinins and/or sugar.

Coding Capacity The amount of protein which can be specified by a given DNA sequence. Estimates of coding capacity usually require assmptios about the absence of introns or of over laping genesas a rough guide 1 megadalton of double-stranded DNA can encode 60-70000 daltons of protein.

Coefficient of Coincidence The quotient of the number of observed double cross-over and the number expected on the assumptions that single cross-over independently. Coincidence varies inversely with interference.

Coefficient of Consanguinity The probability that a related mate is a carrier for a particular trait if the other partner is known to be a carrier for the trait.

Coefficient of Inbreeding The probability that offspring will be homozygous for a given trait if neither the parents not the grandparents are homozygous for that trait and exactly one of the grandparents is a carrier for that gene.

Coelom The body cavity of certain animals, which is lined with cells of mesodermal origin.

Coenzyme B12 Coenzyme derived from Cyanocobalamin. It contains a corrin ring having four pyrrole rings and a central cobalt atom. It is involved as a eoenzyme in methylation reactions, as a methyl carrier and as a prosthetic grouping rearrangement reactions such as the conversion of methylmalonyl coenzyme A to succinyl coenzyme A. It is also involved in the reduction of ribonucleotides to deoxirbonucleotides.

Cofactor A non-protein substance that helps an enzyme to carry out its activity. Cofactors may be cations or organic molecules, known as coenxymes. Unlike enczymes they are in general, stable to heat. When a catalytically active enzyme forms a complex with a cofactor a holoenzyme is produced. An enzyme without its cofactor is termed an apoenzyme.

Cognation Kinship to a common pair of ancestors.

Cohesive Ends Single-stranded termini on DNA molecules that have complementary nucleotide sequences. Compare Blunt Ends.

Coincidence (1) The ratio of observed double cross-over to expected double cross-over calculated on the basis of independent occurrence. This ratio is used as a measure of interference in crossing over. (2) The observed frequency of double cross-over, divided by their calculated or expected frequency. Expressed a pure number; a measure of interference, in positive interference the coincidence s < 1: in negative interference the coincidence is ≯ 1. (3) A measure of the degree of interference in cases of crossing over. It the interference is absolute coincidence is nil; and if there is no inference or even negative interference, the coincidence is maximal.

Cold Antibody (Cold Agglutinin0 Antibodies which agglutinate particles with greater activity below 32 °C. They are lgM antibodies specifically reactive with blood groups1 and in humans, and Raynaud's phenomenon in vivo.

Cold-sensitive Enzymes A group of oligomeri (consisting of several polypeptide chains) enzymes of which about 25 are presently know, which lose their stability and thus enzymatic activity as the temperature is decreased.

Coleorhiza The protective sheath surrounding the radicle in grasses.

Colfactors Plasmid in bacteria conferring ability to produce colicins (antibiotic lipocarbohydrate-proteins).

Coliform Bacteria Cram negative rodshaped bacteria able to obtain energy aerobically or by fermenting sugars to produce acid or acid and gas. Most are founding the verebrate gut, (e.g. Eschericho coli) but some are present in soil, water, or as plant pathogens. Many are pathogenic to man, (e.g. *Salmonella*).

Colinearity (1) A characteristic of the genetic code where by there is a direct correlation in the position of codons on the DNA and the position of amino acids on the protein which the DNA determines. (2) The linear correspondence between the order of amino acids in a polypeptide chain and the corresponding nucleotide sequence in the DNA molecule.

Collateral Inheritance The appearance of characters in some but not all the collateral members of a family.

Collenchyma A specialized type of parenchyma, uullylocated just beneath the epidermis, that functions as supporting tissue. The cell walls are irregularly thickened with cellulose and pectin, the thickening giving distinct patterns to the cells in cross-section. Collenchyma is the first strengthening tissue to be formed in young plants and its able to expand as the young tissues continue development. Compare Schlerenchyma.

Colonisation Factors The pili on enteropathogenic forms of E. coli facilitate adhesion of the bacteria to receptors (probably. GMI gangliosides) on epithelial cells and are often referred to as colonization or adherence factors Colonisation factor antigens may be plasmid coded, are essential for pathoneticity and are strain-specific, for example. K88 (dirrhoes in piglects), CFA/I and CF/II on strains causing similar disease in man.

Colony Hybridization Method used in screening gene libraries held in bacterial host cells. A replica plate from a petri dish of discrete coloniesis blotted on to a filter and, following fixation, in situ hybridization with a radioactive DNA probe is carried out. Autoradiography is then used to identify the colony containing the specific gene and its location on the filter is used to identify it on the original plate.

Colour Inhibitor One of a number of dominant genes that preclude colour formation.

Columella The rounded apex of the sporangiophore within the sporangium of Fungus Class Zygomycetes.

Coloumn Reactor A bioreactor of fermenter where the height is several times the diameter. Typically a column packed with affixed bed is used for immobilized cells or enzymes.

Commercial Sterility The thermal processing of canned foods such that only thermophitic spores survive.

Commercial Sterility The thermal processing of canned foods such that only thermophilicspores survive.

Commisucuum A group consisting of all the organism which are capable of being intercrossed. Communicable Disease A disease that is transmitted directly or indirectly from one person to another.

Communication Action on the part of one organism (or cell) that alters the pattern of behaviour in another organism (or cell) in an adaptive fashion.

Community Any ecologically integrated group of species of microorganisms, plants, and animals inhabiting a given area.

Companion Cell Specialised cell found adjacent to a sieve tube element in some flowering plants.

Comparium A group of individuals within which mating is promiscuous.

Complementary Factors Factors, which individually have similar phenotypic effect but when together cause a different effect. The F_2 segregation show 9.7 ratio.

Complementary Genes (1) Two or more genes, neither of which is capable of producing a phenotypic effect, but which complement each other and work as a team to produce an effect; e.g. at least one dominant allele of genes A,C, and R is necessary to produce colourin maize aleurone. (2) Genes which together produce an effect qualitatively distinct from the effects of any of tem separately. As complementary elements (bateson, Saunders and Punnett, 1906).

Complementation Group A set of mutant alleles exhibiting mutant phenotypic behaviour in the heterozygous state with each other.

Complementation Map An illustration of a series of mutants in a chronomosome segment showing mutually complementing mutants as monoverlapping lines and no complementing mutants as overlapping, continuous lines.

Complementation Test (1) A genetic test to ascertain whether two gene mutations occur in the same functional gene and to establish the limits of the functional gene; also called cis-trans test. (2) The introduction of two mutant chromosomes (sections of chromosomes) into the same cell for the purpose of seeing whether their respective mutations occurred in the same gene.

Complete Digest The treatment of a DNA preparation with a restriction endonuclease for sufficient time that all of the potential target sites within the DNA have been cleaved.

Completely Mixed Bioreactor A stirred tank fermenter or a continous single stage, once trough microbial process that does not involve recycling of the cell biomass. The contents of the bioreactor are mixed by intermittent or continuous stirring of the liquid. This is achieved by mechanical means (using an impeller) or by recycling liquid or passing as gas through it.

Complex Loci DNA sequences that may be interpreted as including more than one structural gene with a common regulatory element. They have been found in a wide range of eukaryotic organism and normally discovered as a series of mutants which behave as a series of mutants which behave as alleles by the cls, trans test, that is they do not complement one another in trans, but they are readily separated by crossing-over, implying a substantial length of DNA. For these latter reasons, these sequences have also been termed pseudoalleles. An example of such a complex locus is the white (w) locus in Drosophila, within which mutations cause loss of pigment from eyes and other body tissues.

Complexity of DNA Summations of the length of different types of sequence in given sample of DNA.

Compound X—chromosomes A group of chromosomes which, in certain forms, acts in place of the usual X-chromosome. This group acts as a compound X during spermatogenesis (e.g. *Ascris incurva* which has 8X + 1Y)

Compound (1) A substance made up of atoms of more than one element. (2) Made up of many units, as the compound eyes of arthropods (as opposed to the simple eye of the same organisms).

Compounds The hetrozygote which contain the unlike members of an allelic series are known as compounds to distinguish them from the other normal heterozygotes. The compounds may show dominance but more often the intermediate type of inheritance.

Concatemer Number of DNA molecules or sequence covalently linked in series. In animal cells, for example, genes coding for ribosomal RNA are often present as multiple, tandemly linked copies and thus form natural concatemers. Components need not however be identical, and during cloning into bacteriophage lambda concatemers of the two phage arms and a DNA insert between them are constructed in *vitro*.

Concatemeric DNA is the substrate for the in vitro packaging of recombinant molecules constructed in cosmid or phage cloning experiments. In such experiments, concatemer formation is favoured by high concentration of both DNA and ligase.

Conditional Lethal Mutation A mutation that kills the affected organism under one set of environmental conditions (restrictive condition) but that is not lethal under another set of conditions (permissive conditions).

Conditioned Medium Cell culture medium that has already been partially used by cells. Although depleted of some components, it is enriched with cell-derived material probably including small amounts of growth will support the growth of cells at much lower density and, mixed with some cloning.

Conditioning (1) A phenotypic alteration resulting from the action of external agents during critical developmental stages. (2) An undefined interaction between media and tissues encouraging the growth of single cell or small aggregates. Conditioned medium may be used to induce growth of cells plated at low densities [below the minimum inoculum size] or cells unable to grow for unknown reasons. Conditioning maybe accomplished by emerging cells or callus contained within a porous material [such as dialysis tubing] into fresh medium for a time dependent on cell densit and a volume related to the amount of fresh medium.

Confluent Cell Culture Culture of cells grown as monolayer on a flat surface, such as a petri dish, in which, by dint of growth and cell replication, the cells form a continuous layer and are all in intimate cell-to-cell contact with one another. Most cells undergo a marked reduction in growth rate when they become confluent, due to contact inibition, but some transformed cell lines continue to grow rapidly and to move over one another following the development of confluence.

Confluent Culture Term applicable to adherent cells only where every cell except those at the very edge are in complete contact with other surrounding cells. By definition, it implies that no part of the substrate remains uncovered by cells.

Conformational Change Alteration in the shape, usually the tertiary structure of a protein, as a result of alteration in the environment (pH, temperature, ionic strength) or the binding of a ligand (to a receptor) or substrate (to an enzyme).

Conjugant (1) A gamete at the time of fertilization (2) One of two pairing chromosomes at synapsis.

Conjugated Protein A protein that contains or is linked to other non-protein moieties; for example metalloproteins (contain metals) lipoproteins (contain lipids) and nucleoproteins (contain nucleic acids).

Conjugation (1) The pairing of gametes in fertilization (2) the pairing of uni-or multicellular organisms in order to consummate the process of fertilization and zygote formation. (3) The chromosomes in the early prophase of meiosis. (4) The process by which DNA is transferred from bacteria of one mating type to bacteria of another during cell-to-cell contact. (5) A process of genetic exchange between bacteria through a physical bridge, mediated by the sex factor. (6) teuniting of two bodies (e.g., synapsed chromosomes in meiosis) or two organisms during sexual reproduction. (7) Joining or pairing in a lengthwise in a lengthwise association of homologous chronsomesin meiosis; the joining of Paramecia an other protozoans prior to fertilization.

Consensus Sequence If a given class of genetic elements e.g., promoters, then the usual form of that sequence is deemed the consensus sequence. Consensus sequences may also be established for genes which encode the protein in different organisms.

Constitutive Heterochromatin Heterochromatin, which is invariably in this conformational state, as is, for example, the chromatin of the centromere regions of chromosomes.

Constitutive An organism is said to be constitutive for the production of an enzyme or other protein if that protein is always produced by the cell under all physiological conditions.

Constriction, Tertiary Any constriction other than the centromere (primary constriction) or one separating a satellite (secondary constriction) from the rest of the chromosome.

Contaminated Area A building area with definite boundaries where hazardous biological work is being carried out, separated from noncontaminated and other contaminated areas by suitable barriers.

Contaminated Service A service or utility, such as water or vacuum, which serves a contaminated area and is therefore segregated from similar services to noncontaminated areas, even though the service itself is non-contaminated. Continental Drift The gradual drifting apart of the world's continents that has occurred over a period of billion years.

Continuity Yates' Correction for A correction applied in the calculation of Normal deviates or X^2's to allow for the discrepancy arising by the observation being discontinuous while tables of the Normal deviated and X^2 are calculated on the supposition of continuity in the variate.

Continuous Cell Recycle Reactor Fermenttion system in which cells are continually removed from the effluent and returned directly to the fermentation vessel. The fermenter may be a simple continuous stirred tank fermenter. The cells are often-separated by centrifugation. Cell recycle maintains a high cell concentration and hence a high volumetric productivity and enables high concentration feeds to be used. A low rate of cell growth may be necessary to maintain a viable culture. The disadvantage of higher capital costs in counterbalanced by improved productivity and reduced equipment size. Several alternatives to a centrifuge for cell harvesting have been investigated. These include simple setting systems where the cells are thermally shocked to halt gas production temporarily an allowed to settle from the fermenter overflow. Alternatively, naturally flocculating organisms may be used. A related technique uses a simple partial recycle fermenter in which the overflow is taken from a vertical pipe rising through the fermenter base and jacketed by a baffled sleeve which tends to reduce the agitation of that region, so allowing the cells to separate partially from the media rising to the overflow nozzle.

Continuous Disinfection Concurrent disinfection: disinfection procedures carried out throughout the course of an illness with careful disposal of contaminated materials.

Conversion (1) The term proposed by Winkler (1930) for a process of interaction between alleles at meiosis. The term was re-introduced by Lindegren (1953) to account for aberrant ratios in the products of meiosis, apparently arising from such interaction. Whitehouse and Hastings (1965) have suggested that if gene conversion is due to correction of mispairing of bases in DNA, it may give rise to reciprocal a well as non-reciprocal recombination. (2) The process by which during recombination, one allele of a gene participating in a cross is lost and is replaced by the other allele. Conversion results from the formation of hetroduplex DNA during recombination and the subsequent repair of noncomplementary bases in the heteroduplex DNA.

Coordinated Biological Insect Pest Suppression A pest population reduction programme involving the directed use of any number of the variable biologically based suppression techniques in compatible combination with the effects of beneficial organisms.

Coordinated Enzyme Repression The process or condition whereby the synthesis of all of the enzymes involved in he biosynthesis of a particular molecule are simultaneously repressed.

Coordinated Enzyme Synthesis Enzymes whose rates of production are observed to vary together. (For example, in E. coli cells growing in the absence of β-galactosidase and β-galactoside permease.]

Core Enzyme RNA polymerase of *E. coli* without the sigma subunit. Together they would be termed the holoenzyme.

Corolla-tube The basal portion of a corolla in which the individual petals are fused to from a tube e.g. Primrose.

Corpora Albicantia Small white bodies arising from corpora lutea when ova liberated are not fertilized.

Cortex (1) (Botany) A primary tissue in roots and stems of vascular plants derived from the corpus meristem that extends inwards from the epidermis to the phloem. It usually consists of parenchuyma cells but other tissues (e.g. collenchyma) may be present, some algae, fungi, mosses and lichens have a well defined region that is termed the cortex although this is different in origin and composition from the cortex of vascular plants. (2) (Zoology) he outermost layer of an organ or part. For example, the outer region of the kidney is called the renal cortex and the surface layer of grey matter in the cerebral hemispheres of the brain is the cerebral cortex.

Cot Unit The product of the concentration of the nucleic acid and time of reaction. It may be expressed in terms of moles of nucleotides times seconds per liter. It is used in the measurement of reassociation of heat-denatured DNA.

Cot Concentration of a DNA sample multiplied by time of incubation in reaction in which strand reassociation is permitted.

Cotransformation The transfer of a well characterised plant trait, expressed in vitro, which is linked to another plant trait or traits not expressed in culture.

Coxsackie Viruses Species of enteroviruses of the picornaviridae first silalted in cozsackie, N.Y. coxsackie A produces diffuse myositis coxsackie B produces focal areas of degeneration in brain and skeletal muscle in mice. Similar to polioviruses in chemical and physical properties. A complication of Cozsackie virus infection can be a persistent myalgia and lassitude which is now recognized as a distinct syndrome (fatigue syndrome).

Crippled Strain Genetically enfeebled strain of a bacterium that will grow only in highly complex media. Such strains are used for any potentially hazardous experiments in gene manipulation to ensure that the bacteria would not survive in the wild; in the case of *E. coli* crippled strains are unable to survive in the human intentional tract.

Critical Night Length In the photoperiodic flowering response of short day plants, the elength of night above which flowering occurs and below which the plant remains vegetative. (The reverse applies in the case of long day plants.)

Critical Orifice An opening through which air passes, the diameter of which is such that the velocity of the air is independent of the pressure differential across the opening, provided the differential is above a certain value. A number of aerosol samplers operate with a critical orifice.

Cro Protein A protein formed in bacterial cells infected with bacteriophage. The cro protein blocks synthesis of a repressor molecule which prevents the lytic cycle of replication occurring.

Crohn's Disease Non-infectious inflammatory bowel disease in the human with numerous environmental trigger factors in addition to an inherited fact.

Cross Products, Sum of Sum of the products of corresponding deviations of two variates from their individual means.

Cross-over Region The segment of a chromosome which lies between any two specified gene loci.

Cross-over Value The sum of the recombination classes in a backcross expressed as a percentage of the total offspring.

Cross-reacting Material (CRM) A defective protein produced by a mutant gene, enzymatically inactive but antigenically similar to the wild-type protein.

Cross A mating or bringing together of genetic material from different genotypes to achieve recombination.

Crossing-Over The mechanism by which linked markers undergo recombination. In general, the term refer to be reciprocal exchange of corresponding segments between two homologous chromatids. However, the reciprocity of crossing-over is problematical in prokaryotes and viruses; and evening eukaryotes, very closely linked markers often recombines by a nonreciprocal mechanism.

Cryobiology Biology of very low temperatures and their effects on living organisms. Cryopreservation is the use of very low temperatures (below-50) to store living material for long periods. Cells (e.g., sperm) may be stored in suitable media such as glycerol, at very low temperatures using liquid nitrogen to be used later for artificial insemination or fertilization. Cryosurgery is the use of ultra low temperatures to freeze and destroy unwanted tissue such as papillomatous warts on the skin.

Cryptic Plasmid A plasmid whose presence causes no discernible change in the phenotype of th ehost.

Cryptic Satellite Satellite DNA which by dint of its base composition, is inseparable from the main band DNA density gradient centrifugation, but detected by other criteria.

Cryptococcus Neogormans Yeast-lie organism: causative agent of cryptococcosis.

Cultivation of Microorganisms, Culture Techniques Deliberate propagationof an organism. Cultivation of an organism may be discontinuous (static) or continuous.

Culture Alteration A persistent change in a cultured cell or tissue anatomyor physiology, including a change in one or more nutritional requirements or a change in its proliferative capacity.

Culture Collection A collection of authentic pure cultures of microorganisms and cultured cell lines.

Culture Media A solid or liquid substance containing nutrient materials for the cultivation of microorganisms.

Culture Tube A test tube used to contain medium and an explant or culture derived from it, usually incubated in a controlled environment.

Culture Vessel A container used to hold medium and an explant or culture derived from it, usually incubated in a controlled environment.

Cumulative Effect The action of two alleles of a gene giving a more pronounced effect than one in the heterozygous condition. Probablyone allele supplies insufficient enzyme for the reaction conditioned by two. The hybrids are distinguishable from parents. The effect is sometimes referred to as incomplete dominance.

Curing Agent A chemical catalyst added to the components of a polymer to effect polymerization.

Curly-Lobe-Plum Drosophila A balanced lethal system in chromosome 2 in Drosophila.

Cushing's Syndrome A type of hypertensive disease in humans due probably to the over secretion of cortisol due in turn to excessive secretion of adrenocorticotrophin (ACTH). Adrenal tumorous are the usual primary cause.

Cyanobacteria (Cyanophyta) Group of bacteria, previously known as blue-green algae. They are unicellular and non-motile, often coated with mucilaginous slime, and the photosynthetic membranes occur as thylakoid within the cell cytoplasm rather than being in a membrane-bounded chloroplast.

Cyanocobalamin (Vitamin B12) One of the water-soluble B-group of vitamins. It has a complete organic ring structure at the centre of which is a single cobalt atom. Foods of animal origin are the only important dietary source. A deficiency in man leds to the development of pernicious anaemia since the vitamin is required for the development of red blood cells.

Cyanogen Bromide (CNBr) Chemical agent for cleaving Peptide Bonds in polypeptide chains. CNBr act specifically at methionine residues converting the free °C-terminu to homoserine lactone.

Cybrid A cell or plant cytoplasmic (heteroplast0 with the nucleus of one and cytoplasmic organelles of another, or of both cells or plants.

Cyclic AMP Adenosine monophosphate with phosphate group bounded internally (phosphodiester bound between 3' and 5' carbon atoms) to form cyclic molecule, Active in regulation of gene expression in bacterial and eucarytoic cells.

Cylic GMP (cyclic guanosine Triphosphate) Energy-rich molecule which is suspected of being the transmitter of the signal form photosensitized Rhodopsin to the plasma membrane during photoreception in the vertebrate eye. Compare Cyclic Amp.

Cytoplasmic Incompatibility "A kind of prefabricated sterility" in the matings between certain allopatric populations of the same species, resulting from the inability of the sperm from one string to penetrate the cytoplasm and fertilize the egg of the other string.

Cytoplasmic Male Sterility A type of pollen sterility transmitted through the plasma, materally inherited.

Cytosol The fluid portion of the cytoplasm, excluding organelles and other solids.

Cytotaxonomy The use of chromosome number, size and space in the classification of organisms.

Cytotoxic Drug Any drug that kills cells. The term is sometimes used specifically to refer to compounds that kill cancer cells.

Cytotoxic Reaction Allerigic reaction directed against antigens on the cell surface.

D

D Cellos (d cells) Cells of the pancrease; about 5% of the cells present in primate pancreas with small argentaffin-positive granules. Their function is unclear, but they may release somatostatin.

D Loop (Displacement Loop) Structure produced in replicating DNA when a 'bubble' or 'eye' of DNA is formed and can be visualized in the electron microscope. The production of the loop results from the DNA being double stranded on one side of the bubble and single-stranded on the other and it is exploited as a technique to demonstrate hybridization of an exogenous sequence to one of the complementary strands, leaving the other displaced strand looping out from the helix.

D-value The decimal reduction time. The sterlization time necessary to reduce a microbial population by 90% (one log). It must be assumed that the rate of kill is linear with respect to the logarithm of the surviving population, then:

Dalton The unit of atomic mass: one dalton is exactly ½ the mass of a single neutral atom of ^{12}C, the most common isotope of carbon.

Dam An E. coli mutant which is deficient in the methylation of its own DNA. The methylation of DNA in dam+ strains modifies the DNA so that it is no longer recognized by certain restriction enzymes, e.g. BclI.

DAP (Diaminopimelic acid) A seven carbon diamino acid that occurs as a component of cell-wall peptidoglycan in some bacteria.

Dark Reaction The reactions in photosynthesis which occur after NADPH and ATP production, and which take place in the stroma of the chloroplast; constituted the Calvin-Benson cycle. By means of the reaction, CO_2 is incorporated into carbohydrate.

Dark-field Microscopy A type of microscopic examination in which the microscopic field is dark and any objects, such as organism, are brighdy illuminated.

Darwinian Fitness The relative fitness of one genotype compared with another, as determined by its relative contribution to the following generations.

Data backup A copy, or the process of copying data stored on magnetic storage devices (e.g. Floppy disk) to ensure security in the event of corruption.

Data Compression The process of converting data to a more compact representation that occupies less disk or memory space.

Data Reduction The process of reducing the number of recorded values from an observation without significant loss of important information.

Data Structure Programming languages as Fortran. Basic and Pascal provide for a small set of fundamental types of data representation such as integer and real numbers, characters, arrays logical or Boolean values. These simple data Types can be combined by the programmer in a variety of ways (depending on the programming language) to form data structures that can be manipulated as single data types and used to represent the required information. In Pascal such complex data structures are called records.

Data Terminal Equipment Any mechanical or electrical device capable of displaying or printing output from a computer system.

Dating Technique A method that examines the biological, chemical or physical characteristic of matter in order to determine its geological age. With biological material, the technique of carbon dating is often used.

Dauer Modification (1) A lasting heritable change, presumably cytoplasmic produced by treatment (2) A phenomenon in which environmentally produced modification of an organism reappears in a lessening degree for one or more subsequent generations.

Daughter In microbiology or molecular biology, one of the products arising from division of a cell or replication of a molecule of DNA. In such processes, two individuals or molecules will be produced which are genetically identical to the parent cell or molecule.

Death Curve A graph which shows the rate at which a microbial population dies. It can be plotted as a logarithm of the number of survivors at any given time. The curve shows that death occurs as a first order reaction.

Debranching Enzyme An enzyme that hydrolyzes the (1, 6) links found in amylopectin and similar polsaccharides.

Decay Accelerating Factor Plasma protein which regulates complement cascade by blocking the formation of the C3bBb complex (the C3 convertase of the alternate pathway). Widely distributed in tissue but deficient in paroxysmal noturnalhaemoglobinuria.

Defective Virus A virus that by itself is unable to reproduce itself when infecting its host, but that can grown in the presence of another virus.

Defense Guild A group of plant species that interact with each other in their relationships with predators; the interactions are usually chemical.

Deficiency Disease A disease resulting from a diet that lacks an essential nutrient, such as a vitamin, essential amino acid or trace element.

Deficiency Loop The loop of the nonsynapsing portion of an unaltered chromosome caused by adeficiencynthechromosome caused by a deficiency in the homologus chromosome.

Definitive Nuleus A nucleus formed two polar nuclei. The definative nucleus later fuses with a male nucleus to form the triploid endosperm nucleus.

Dehumidifier An apparatus that removes moisture from the air.

Dehydrogenase An enzyme which oxidizes a substrate by removing hydrogen atoms from it.

Dehydrogenation The usual type of biologic oxidation that occurs in microorganisms. The enzyme dehydrogenase causes the hydrogen do not lose part of its hydrogen, and as hydrogen and electrons are transferred energy is released.

Delay Hypersensitivity An allergic or hypersensitive reaction to a foreign substance resulting from a cellular immunity to that substance. It is T cell mediated hypersensitivity in which the reaction reaches the maximum in about 24 to 48 hours after exposure to the antigen that elicited the original hypersensitive state.

Deletion Method Procedure for isolating Messenger RNA (mRNA) specific to a particular gene, involving Hbridization of total cellular mRNA) specific to a particular gene, involving Hybridization of total cellular mRNA to the DNA from a particular cell line, under conditions favouring such hybridization. When the DNA is derived from a cell line with a deletion at a particular locus, the population of mRNA that fails to hybridize is likely to be specific for the gene or genes involved in the deletion.

Delta-haemoglobin (HbA_2) Comprises between 1, 5 and 3. 2% of the haemoglobin in healthy adult humans. HbA_2 has two alphaglobin chains, giving an $A_{2\text{-}2}$ formula. The deltaglobulin gene is found close to the betaglobin gene on chromosome 11.

Denatured DNA DNA that has been converted from double-stranded to single-stranded to single-stranded form by breaking the hydrogen bonds joining the two complementary strands.

Denatured Protein A protein that has lost its natural configuration by exposure to a destabilizing agent such as heat.

Dental Plaque An aggregation of bacteria and organic material on the tooth surface.

Deoxyadenlic Acid The deoxyribonucleotide containing the purine adenine.

Deoxycholate Detergent Relatively mild, anionic detergent that causes minimal unfolding of the protein molecule and can therefore be used to isolate membrane protein without impairing their activity. Detergents are small Amphipathic molecule that can be used to isolate membrane proteins.

Deoxycytidylic Acid The deoxyribonucleotide containing the purimidine cytosine.

Deoxyribonucleic Acid DNA The genetic molecule of most organisms, except some viruses. Double stranded polymer of nucleotides arranged along a deoxyribose and phosphate backbone. Structure was proposed by James Watson and Francis Crick in 1953.

Deoxyribose A live-carbon sugar having one oxygen atom fewer than the parent sugar, ribose; a component of DNA.

Depolarizaton Loss of resting potential, especially in synaptic membranes and motor end plates of neurons, this is induced in the neuroneby Acetylcholine, which induces movement of sodium and potassium ions into and out of the membrane.

Depression Displacement of a repressor protein from apromoter region of DNA. Hen attached to the DNA, the repressor protein prevents RNA polymerase from initiating transcription.

Derivative The first derivative of a list of data describing a continuously varying observation is the first differential of the mathematical function the describes the observations. In other words, the first derivative is a measure of the local slope (the rate of change) in the observed values. It is generally calculated for several point either side of the value being considered. Values of zero in the first derivative indicate the position of peaks or troughs in the observed data.

Dematophytes Fungi which infect superficial tissues, the skin, hair, and nails.

Desiccant A chemical compound, such as calcium chloride, sodium hydrzide, concentrated sulphuric acid, silica gel, alumina gel or molecular sieves, used to remove water from a sample or to keep the air in a container dry, many of the complex reagents and chemical compounds used in biochemistry are subject to decomposition by hydrolysis. Hence, they are stored in the presence of a desiccant in order to reduce such loss to a minimum.

Desoxycorticosterone A steroid produced by the suprarenal cortex which has a progesterone-like action and possibly androgenic properties also.

Desynapsis The falling apart of synapsing chromosomes before metaphase (synapsis having girt occurred during the meiotic prophase).

Detarchment The separation of a single X-chromosome from attached X-chromosoes, as a result of crossing over between the X and Y-chromosome near the centromere.

Detassel To remove tassels from corn plants to accomplish cross fertilization on a field scale, special rows (usually one in five) are planted to supply pollen for detasseled rows.

Detention Time The length of time a liquid or solid component of the mixed liquor in a fermentation process remains in the fermenter or bioreacter.

Detergent A cleaning agent; one of numerous synthetic cleaning preparations chemically different from soap. Detergents are commonly used to prepare work areas for aseptic plant tissue culture manipulations and to remove dirt and microorganisms from plant material prior to explanation.

Determinant An element in the cell postulated to control heredity and development. The precursor of the gene concept.

Determinate Growth Growth of limited, often set duration of fixed fate; as in most floral meristems and leaves. The direction of differentiation becomes irreversibly established, this process contrasts with the usual culture growth which is (theoretically) infinite, and so indeterminate.

Determination or topophysis The process of commitment to a specific pathway of development [cell, tissue, organism, growth form]. This commitment may occur well in advance of the appearance of the phenotype. The perpetuation of the phenotype provides an example of the stability of this process.

Deuterostome One of two major lines of evolution in animals, characterized by radial cleavage, enterocoelous development, and other traits.

Development Genetics The study of the operation of genes during development.

Devort Test More sensitive version of the Ames test, used for detection of potential carcinogens of relatively weak toxicity. Test assay uses lambda Phage lysis of previously lysogenic *E. coli.*

Dexamethasome Glucocorticoid hormone like drug often used clinically for suppression of inflammation and inhibition of Acth secretion in adrenal hyperplasia Dextran. A polysaccharide (glucose plymer) produced by a wide range of microorganisms, sometimes in large amounts.

dG Deoxyguanosine One of the four nucleoside constituents of DNA. For structure.

DHFR Dihydrofolate Reductase An enzyme involved in the synthesis of prunes and pyrimidines. It is inhibited by trimethoprime in prokaryotes or by methotrexate in prokaryotes and eukaryotes, genes for certain methotrexate-resistant DHFRs can be used as selective markers on cloning vectors for prokaryotic and eukaryotic cells.

Diad Two daughter cells arising from aspore mother-cell, as a result of meiotic irregularities, instead of the tetrad. (2) An univalent chromosme at meiosis.

Diagnostic Ensyme An enzyme used as a specific diagnostic agent in order to obtain a qualitative or quantitative indication of the presence of a given compound which serves as the enzyme substrate.

Diagnostic Kit A package that contains all the required reagents in suitable form, to carry out one or more specific assays which may be based on immune, chemical or enzyme-catalyzed reactions. To be accepted for use in health care, such procedures and reagents require validation and acceptance by health authorities.

Diagnostic Procedure A test or assay used to determine the presence of.an organism, substance, or nucleic acid sequence alteration that represents a disease or pathogenic condition.

Diagnostic Reagent A chemical, buffer enzyme, antibody preparation, etc., that issued for the purposes of diagnostic kit.

Diakinesis The term coined by hacker (1897) for the stage of late prophase 1 of meiosis when the chromosomes are well separated from one another (GK), kinesis = movement; dia =apart). This stage is recognized by the highly condensed condition of the chromosomes, the homologus pairs of which are held together by chiasmata.

Dialysis Fermentation A technique in which the cells are retained within a fermenter using a membrane at some point.

Diandric Of a genetic character; passing from female to female through the male bymeans of the sex chromosome (birds and moths).

Diapause State f a reduced rate of Metabolism which can form part of life cycle of holometabolous insects, and in which development temporarily ceases.

Diarrhea Abnormal fluidity of fecal discharge.

Diaster The stage at the end of mitosis or meiosis when the daughter chromosomes are grouped near the poles of the spindle.

Diatomite Silica-containing shells resulting form centuries of growth of diatoms.

Dibutyrl Cyclic AMP An analogue of cyclic AMP which shares some of the pharmacological effects of this nucleotide but is generally believed to entre cells more readily on account of its greater hydrophobicity.

Dichlamidius Chimaera A periclinal chimaera in which the genetically distinct periperal tissue is two layers of cells thick.

Dichotomous Outcomes An experiment with two outcomes, generally referred to as Success and failure.

Dictyokinesis The division of the golgi apparatus at mitosis and the even distribution of dictyosomes between the daughter cells.

Differential Affinity The failure of two chromosome to pait at meiosis in the presence of a third, although they pair in its absence.

Differential and Interference Distances these are, respectively the distances between the centromere and the mean position of the proximal chiasma, an mean distance between the proximal or first and the next proximal or second chiasma.

Differential Interface Microscope Type of light Microscope used to study living cells by shifting the phase of light passing through a specimen and thus producing sharp contrast.

Differential Reproduction Reproduction in which different types do not contribute to the next generation in proportion to their numbers.

Diffusion Potential Maintenance by the balance between ionic pumping ad ionic diffusion which results in electrochemical gradients across cell membranes.

Diffusion Movement of small molecules in the liquids and gases as a result of the random tracking of the molecules themselves. It tends to result in movement away from molecular concentration throughout a system.

Digest To convert complex, usually insoluble materials to simple often soluble units through enzymatic action; as in the removal of plant cell walls during protoplast formation. The process is digestion.

Dihybrid An individual heterozygous for two pairs of alleles; also said of a cross between individuals differing in two gene pairs.

DihydrofolateReductase Enzyme that ctalyse the NADPH-dependent reduction of folate totetrahydrofolate, an essentialcofactor in the synthesis of glycin, purines and thymidine.

Dikaryon A mcellium in whch each cell contains two genetically dissimilar nuclei. Dikaryotic mycelium is extensive in Basidiomycetes and forms the entire fruiting body (mucelium is extensive in

Basidiomycetes and forms the entire fruiting body (mushroom). It also leads to the formation of basidia and basidiospores. In Ascomycetes, only the ascogneous hypae are dikaryotic.

Dikaryosis Condition of being a Dikaryon. It may result from artificially induced cell fusion, but occurs naturally in the sexual cycle of some fungi.

Diluting Factor A minor factor which, by itself, has no effect but which lessens the effect of major factor.

Dikaryon A mecellium in which each cell contains two genetically dissimilar nuclei. Dikaryotic mycelium is extensive in Basidiomycetes and forms the entire fruiting body (mushroom). It also leads to the formation of basidia and basidiospores. In Ascomycetes, only the ascogenous hyphae are dikaryotic.

Diluting Factor A minor factor which by itself, has no effect but which lessens the effect of major factor.

Dimethylsulphoxide (DMSD) Compound that destabilize and denatures the double helix of DNA, DMSO has been used to 'induce' activity of specific genes, previously unexpressed, in tissue culture cells.

Dimorphic Occuring in two forms; characteristic o certain fungi, growing in the yeast phase under one set of conditions and filamentous under other conditions.

Dimorphism The production of two distinct types of flower on the one plant. (2) The occurrence of two distinct forms within a species Polymorphism.

Dioecy, haploid or Diploid The condition in which male and female gamates are borne on separate haploid or diploid individuals respectively. Has usually been termed uniseuality in animals.

Diphyetic Acid A compound found by a symbol designating a particular chromosome, indicates an individual in whose somatic cells both members of this particular archromosome-pair are present. Diplo-is used to distinguish individuals of this type from haplo- and triplo-individuals.

Diploid Cell Strain Cultures of cells or tissues derived out of primary cultures

Diploid Parthenogenesis In diploid parthenogenesis the reduction division of meiosis is abortive so that diploid ova are formed, which later, may develop without fertilization.

Direct Pest An organism which causes immediate and direct damage to a maketable item, such as a fruit, even at low population density.

Direct Repeat Sequence Identical or near identical sequence of DNA present as two or more copies in the same orientation in the same molecule

Directed Mutation Selective mutation of a particular gene sequence or class of sequence. Such mutations maybe achieved by the use of drugs which have a particular affinity for the sequence in question (e.g. active as compared to inactive gene sequences) or which tend to alter permanently the a stage of differentiation of cells in culture.

Disease Garden A nursery in which plants can be grown under optimum conditions for any or several particular disease as a means of studying resistance.

Disease Resistance The ability to resist disease or the agent of disease and so remain healthy. Resistance or tolerance to disease is topic of intensive interest in plant tissue culture work. Screening or selection at the cellular level is followed by plant regeneration and screening of these and their progeny.

Disease A physiological or anatomical plant disorder or disturbance of biotic or abiotic origin that interferes with its structure, function or value: as some plants suffer from nutritional deficiency diseases in culture.

Dissogony Having two periods of sexual maturity during the life cycle one during the larval stage and the other during adulthood.

Dissolved Oxygen The amount of oxygen, expressed in terms of equivalent volume under standard conditions, moles, or weight, in true solution in unit volume of culture medium or other liquid.

Distinguishable Hybrid A hybrid in which intermediate inheritance is expressed (i.e., the heterozygous combination is distinguishable by a phenotype.).

Distributed Array Processor A parallel computer architecture available in some ICL mainframe computers and used for fast calculations on matrices.

Dizygotic Twins Fraternal twin produced by the fertilization of two separate eggs by two different spermatozoa.

DLVO Theory Theory of colloid flocculation advanced independently by Derjaguin and Landau and by Vervey and Overbeek and subsequently applied to cell adhesion.

DMSO (Dimethyl Sulphoxide) Much used as a solvent for substances which do not dissolve easily in water and which are to be applied to cells (for example cytochalasin B, formyl peptides), alsoas a cryoproectant when freezing cells for storage. It is used lclinically for the treatment of arthrit although its efficiency is disputed.

DNA Amplification Fingerprinting (DAF) A method of general DNA amplification using a single primer of between 6 and 8 nucleotides in length. The primer is arbitrarily chosen and may generate as many as 80 amplification products, which can be removed by a variety of methods, including polyacrylamide gel electrophoresis (page) and silver staining, agarose electrophoresis, and automated analysis using either DNA sequences (using fluorescent primers and laser detection) or capillary chromatography. Method as developed in 1990 by the University of Tennessee and is used to distinguish organism (e.g. cultivar of one crop species), gene mapping and diagnostics.

DNA Cloning Technique whereby DNA from the source is inserted into a cloning vector and then introduced into a host bacterium. Loose ends are reigated using DNA ligase to produce a coherent recombination DNA molecule. Recombinant DNA molecules are introduced into a host bacterium (usually . *coli*) by transformation if a plasmid vector is used. Or if the cloning vector is bactriophage based by transfection or infection (following *in vitro* packaging).

DNA Construct A DNA molecule inserted into a cloning vector, usually a plasmid.

DNA Filter Assay An analytical procedure used to recognize recombinant DNA in cloned cells. The transformed cells only represent a small percentage of the total population of cells present in a culture. The culture is diluted and platedon petri dises at such a concentration that individual colonies form from each cell. A replicate of the pattern of cells on the plate is taken using a cellulose intrate filter. The cells are lyzed, and the filter heated to bind the DNA onto the filter in adenatured form in a position equivalent to that of the cell colonies on the petri dish. The filter is then treated with the probe which combines by DNA hybridization only at the positions of transformed DNA.

DNA Fingerprinting Technique in which the banding pattern of DNA fragments is compared and can be used in many species, including the human, to indicate related-ness. DNA is digested with restriction enzymes, run on an electrophoretic gel and blots made from the gel. Such biots are then hybridized with radio labelled probe DNA consisting of cloned sequences of the short interspersed repeat sequences (such as the ALU sequences in the human) which are common in vertebrate genomes autoradiography then provides a distinct banding pattern of DNA fragments containing such DNA sequences.

DNA Gyrase Enzyme that increases or introduces positive Super coiling in covalently closed circular DNA molecules generated by DNA replication. This achieved by breakage of at least one DNA strand followed by untwisting and subsequent resealing of the duplex. DNA gyrase is an alternative term for a 2Topolsomerase.

DNA hybridization Production of double stranded DNA or DNA –RNA hybrid molecules from single-stranded molecules from different sources by the formation of hydrogenbonds performed as a mixed phase with one of the nucleic acid strands immobilized to a solid support. This technology has been used extensively to study gene structure, genome organization and the control of gene expression. Normally a fragment of DNA is radio labelled and then used as a hybridization 'probe' to detect homologus or nearly homologus sequence that have been immobilized onto a solid supporting matrix (i.e. by Southern Blotting). Hybrid formation is a complex reaction depending on the type of probe used. Using a single stranded probe the rate of hybridization follows first-order kinetics and is affected by probe strand length, complexity, ionic strength, temperature, pH and viscosity. Annealing of nucleic acid strands to form hybrids is a reversible process. The stability of perfectly or imperfectly matched duplexes depends on clearly defined criteria. Hence sequence relatedness of hybrid duplexes can be determined by changing these criteria in a procedure known as Stringent Washing.

DNA Library (DNA Bank) Collection of cloned DNA fragments that represents part or all of the genome of an organism. Most libraries are partial simply as a result of the method of formation and the comparative rarity of some gene sequences. Genomic and cDNA libraries differ in that the former comprises pieces of genomic DNA and therefore including introns and noncoding sequences, whereas the latter being complementary to mRNA, contain only coding sequences.

DNA Ligase An enzyme catalyzing the covalent union of segments of an interrupted sugar-phosphate strand in double-stranded DNA.

DNA Melting Denaturation of double-stranded DNA to yield single-stranded DNA. Melting, which can be induced by increased temperature when DNA is dissolved in a suitable salt solution, results in a reduction of viscosity and an increase in ultraviolant absorption. This increase in light absorption is the hyperchrome effect. Melting can be followed by annealing or renaturation, which involves specific hybridization between sister strands and is used in the derivative of Cot Plots.

DNA Modification Alternation of DNA by the addition of various side groups (often methyl groups) brought about by specific enzymes.

DNA Polymerase Anyone of a system of enzymes that catalyze the formation of DNA from deoxyribonucleotides, using one strand of DNA as a template. DNA Polymerase I. Holoenyme. It has 5' exonuclease activity in addition to DNA synthesis. This bifunctional activity enables the enzyme to use nicks or gaps in double stranded

DNA as a starting point for DNA synthesis. The 5'-exonuclease activity degrades that DNA strand which is complementary to the template strand and thus forming nick. DNA synthesis beings at 3'-end of nick and produces a new strand of DNA complementary to the template. The net result is the movement of nick along the template strand (nick translation) until all the DNA complementary to the template strand (starting from the site of the origin of nick to the 5'-end of the template strand) is replaced. Applications cf DNA Poly I: (1) Enzyme is used with radioactive or biotinylated nucleotides to prepare labeled DNA of high specific activity. (2) DNA polymerase can also catalyze de novo DNA synthesis.

DNA Polymerase 1, Klenow Fragment Treatment of DNA polymerase I holoenzyme with protease results in production of two protein fragments. The larger fragment, also known as Klenow fragment does not exhibits 5'- exonuclease activity (5'-exonuclease activity is exhibited by intact enzyme). This Klenow fragment issued to synthesize DNA when there is no need of removing the DNA strand which is complementary to the template strand. Application of kleno Fragment (1) in DNA sequencing by dideoxy-method. (2) For production of second strand of cDNA. (3) Radio labelling by filling in 5'-single stranded extension on double stranded DNA. (4) Mutagenesis of DNA with synthetic oligonucleotides. (5) in labeling the DNA by random primer method.

T4DNA Polymerase T4 DNA polymerase lacks 5' exonuclease activity but has a very active 3'-exonuclease action. This allows radio labeling of DNA fragment by replacement synthesis. In this method it acts on both the 3'-end of double stranded DNA resulting in long 5'-single stranded extensions. DNA complementary to these single strands are synthesixed with deoxynucleoside triphosphase enzyme in the presence of radio labelled compounds. Applications of T4 DNA Polymerase (1) Useful in generating 5'-single stranded ends. (2) The enzyme can also be used in radio labelling of DNA. Taq DNA Polymerase. Taq DNA polymerase is isolated from the thermophilic bacterium thermos aquaticus. This enzyme has the highest DNA polymerase activity at a pH of 9 and temperature around 75 °C DNA polymerase activity is resistant to incubation at high temperatures like 95 °C. this enzyme consists of a single polypeptide chain with a molecular weight of 95,000. It lacks 5'to 3' and 3'to5' exonuclease activities. The highly thermostable DNA polymerase from Thermus aquaticus Ta1) is ideal for both manual and automated DNA sequencing because it is fast, highly progressive, has little or no 3'- exonuclease activity and is active over a broad range of temperatures. Applications of Taq Polymerase (1) It can be used in DNA sequencing studies.

DNA Probe A short length of DNA of known sequence or origin that is labeled in such as way that it can be subsequently recognized. The label may be an isotropic marker, such as radioactive phosphate, or a chemical compound that can act as a binding sit for an affinity label. The probe is used in DNA filter assay.

DNA Rearrangement Wholesale movement of sequences from one position to another in DNA, such as occur somatically for example in the generation of antibody diversity.

DNA Repair Enzyme correction of errors in DNA structure and sequence which protect genetic information against environmental damage and replication errors.

DNA Restriction Endonuclease Most important class of Restriction Endonucleases, class 2, which cut double-stranded DNA molecule only at sites characterized by a specific nucleotide sequence. Restriction enzymes are isolated from bacterial cells, and are tools for molecular biologists. Several hundred restriction enzymes are now known each with a specific sequence requirement dictating where it will cut DNA. Most recognize sequences between four and six base pairs long, which in turn are likely to occur by chance once every few hundred to once every few hundred to once every few thousand nucleotides along a DNA molecule.

DNA Topoisomerase An enzyme capable of altering the linking number of DNA. Usually measured by following a change in the degree of super coiling of double-stranded DNA molecules. Various topoisomerases can increase or relax supercoiling, convert single-stranded and duplex unit knots in single-stranded and duplex rings, catenate and decatenate duplex rings. Topoismerase2 of E. coli. = gyrase.

DNA Transfecton A technique originally developed to allow viral infection of animal cells by uptake of purified viral DNA rather than by intact virus particles. Term is used to describe application of same methodology for introduction of DNA into cells, such as active oncogene from tumours into tissue culture cells. DNA tumour virus Virus with DNA genome which can cause tumours in animals. Examples are papovaviridae, adenoviridae and Epstein Barr virus.

DNA Tumour Virus Virus with DNA genome which can cause tumours in animals. Examples are papovaviridae, adenoviridae and Epstein Barr virus.

DNA Virus A virus in which the nucleic acid is double or single-stranded DNA (rather than RNA). Major groups of double-stranded DNA viruses are papovaviridae. Adenoviridae, herpetoviridae, large bacteriophages and poxiviruses; of single-stranded, parvoviridae and coliphages QX174 and M13.

DNA, Native Double-stranded, non denatured DNA in a preparation that has been extracted from a particular known biological source-viral, bacterial, or eukaryotic tissue as distinct from any additional material added as a probe or marker.

DNA-Protein Interaction Molecular interaction that is a crucial aspect of living system both to form stable complexes such as Nucleosomes, and transient ones involving RNA Polymerase and DNA Polymerase and gene regulatory proteins.

DNA-Driven Reaction Reaction in which nucleic acid hybridistion is utilized in an experimental situation, and complementary DNA is greatly in excess of a radioactive tracer RNA. The RNA I thus forced to complete with a much greater proportion of hybridization DNA and thus will bind to highly repetitious DNA sequences.

DNA-RNA Hybrid A double helix that consists of one chaon of DNA Hydrogen bonded to a chain of RNA by means of complementary base pairs.

DNA Deoxyribose nucleic acid A special form of nucleic acid which constitutes the hereditary substance of the chromosmes. Its detailed structure consists of a double helix as proposed by Watson and crick. A double-stranded, helically coiled nucleic acid molecule that is the genetic material of living organism and many viruses.

DNP A commonly used hapten.

Dilipore The circular perforation found in the septa of basidiomycete hyphae. They have a characteristically thickened edge which is often associated with endoplasmic reticulum.

Dominance, Hypothesis of Antithetical In hybrids between externally diverse parents, natural selection will tend to encourage those modifiers favouring one parental extreme or the other and suppressing intermediates.

Dominance The relationship of two allelomorphs where the single gene heterozygote resembles one of the two homozy-goteresembles one of the two homozygous parents rather than the other on an arbitrary scale distinguishing the two phenotypes.

Dominigenes Modifying genes which are able to modify the dominance of another gene.

Donor Parent That parent from which by backcrossing one or more genes are transferred to the backcross parent.

Donor plant (mother plant) The source plant used for propagation, whether an explant, graft or cutting . Source plants used for micropropagation are usually pathogen-tested (disease-free).

Dosage Compensation Any mechanism by which effective dosages of sex-linked genes in organisms with an XX-XY or XX-XO mechanism of sex determination are made equal.

Double Fertilization The union of the second sperm nucleus with the primary endosperm nucleus with the primary endosperm nucleus to form the fusion nucleus, which gives rise to the endosperm tissue.

Double Helix DNA exists in the cell as a double stranded structure coiled in the form of a helix

Double Stable Transformant It is stably transformed for two antibiotic resistance genes.

Double Unstable Transformant It is unstably transformed for two antibiotic resistance genes.

Doubling Time Time taken for a cell to proceed through the cell cycle and divide by mitosis, or for a complete culture to undergo a doubling of cell number.

Downstream (1) In molecular biology, the stretch of nucleotides of DNA that lie in te 3' direction of the site of initiation of transcription, which is designated as + 1. Downstream nucleotides are marked with plus signs, e.g., +2. + 2 10. Also to the 3' side of a particular gene or sequence of nucleotides. (2) In chemical engineering, those phases of a manufacturing process that follow the biotransformation stage. Refers to recovery and purification of the product of a fermentation process. Also called downstream processing.

Driver DNA Excess DNA used in a DNA-Driven reaction.

Droplet Infection Infection acquired from recently contaminated air via suspended droplets.

Droplet An airborne particle consisting primarily of liquid. Droplets are usually generated by talking, sneezing, or coughing, and while some settle out almost immediately, many dry to become droplet nuclei to add significant numbers of microorganisms to the air.

Drug Resistance Gene A gene that codes for an enzyme which enables the host ell to resist the effect of a drug.

Dry Cell Weight The weight f cells contained in an aliquot obtained after drying to constant weight in an oven at 105 °C. Dry weight are often expressed in terms of grams of material per litre for cubicmetre of fermenter.

Dry Heat Sterilization Thermal sterilization carried out in the absence of added moisture. For the same sterilization time, dry heat usually requires a higher temperature than moist heat to achieve the same degree of sterilization.

Dry Ice Frozen [solid] carbon dioxide [CO_2] It is commonly used as a refrigerant.

Dt Gene Maize A genetic mutagen in chromosome 9 in maize, causing a (chromosome 3) to mutant to A, yielding dotted kernels or leaves streaked with an-thocyanin.

DTH (Delayed Type Hypersensitivity) This term includes the delayed skin reactions associated with type 4 hypersensitivity.

Dual Culture A culture system that includes plant tissue and one organism (such as nematode species) or microorganism [such as a fungus]. Dual cultures are used to study host-parasite interactions or in the production of axenix cultures for a variety of purposes. Usually the microorganism selected is an obligate parasite.

Duplex DNA DNA in the double stranded form forming a double helix as described in the Watson-crick model that is with the two polynucleotide chains lying in opposite 3'-to5'-polarities, intertwined and hydrogen-bonded together.

Dynein Calcium and magnesium-activated ATPase located in the dynein arms which are attached to subfire A of the microtubule doublet of cilia.Dysgenic Tending to be harmful to the heredity qualities of a species.

Dysinetoplasty Absence of an organized kinetoplasty froma flagellate protozoan cell.

Dysploidion A species composed of the morphologically similar members of a dysploid series, the individuals of which are sexually reproductive.

Dysteleology The supposition that nature lacks any foreordained direction or purpose.

E

E Classification (enzyme Classification) Classification of enzymes based on the recommendations of the committee n enzymes nomenclature of the international union of biochemistry. The first number indicates the broad type of enzyme (1 = oxidoreductase; 2 = transferase; 3 = hydrolase; 4 = lyase; 5 = isomerase; 6 = ligase (synthethetase). The second and third numbers indicate subsidiary grouping and the last number, which is unique is assigned arbitrarily in numerical .order by the committee.

Eadie Plot A method of treating date from investigations of enzyme kinetics in order to obtain straight line plots from which the various kinetic constants can be calculated in such plots, the value obtained by dividing the substrate concentration. This gives a slope with a value equal to the reciprocal of the maximum velocity and an intercept on the X axis equal to the negative of the K_m.

Early Gene Genes contained in the circular genomes of viruses such as Bacteriohage T4 and SV40. Early genes are transcribed early in the viral replicative cycle, and some of their products are necessary for the transcription or translation of other genes in the genome, termed late genes in double –stranded DNA viruses, the early genes tend to be clustered on one DNA strand and late genes on the other; transcription involves both strands, each of which is read in an opposite direction.

Early Proteins Protenaceous substances, particularly enzymes, synthesized during the early stages of bacteriophage multiplication: are involved in various biochemical activities during phage multiplication.

Ecdysis (1) Periodic moulting or shedding of the exoskeleton in

Arthropoda to permit growth. In insecta and crstacea, it is controlled by Ecdysone. (2) Periodic shedding of the epidermal layer of skin in reptilia.

Echinoderm A member of the pylum Echinodermata, such as a sea star or sea urchin.

Echo Viruses Any syntheti or natural chemical of plant or animal origin exhibiting biological activity similar to the true molting hormone, ecdysone.

Eclipse (Greek Eklipsis, failure to leave to come out). The period during bacteriophage multiplication which starts with the injection of nucleic acid into the host cell till the formation of first complete phage particle.

Eco Ri Restriction Enzyme Type 2 restriction endonuclease (DNA Restriction endonuclease) isolated from E. coli that recognizes the hexanucleotde target sequence: 5' GAATTC 3' 3' CTTAAG 5' Gleavage of the DNA duplex occurs at positions four nucleotide apart in the recognition sequence to yield fragments of DNA with protruding cohesive 5'- termini four bases in length.

Ecophene The range of phenotype produced by one genotype within the limits of habitat under which it is founding nature. One genotype must be taken to mean a class of closely similar genotypes.

Ectopic Expression Expression of a molecule in an abnormal site. An example is the production of hormones by some tumours, including secondary tumours.

Ectosomes Specific cytoplasmic granules characteristic of the primordial germ cells and stem-cells in copepods. A from of 'germ-cell determinant.

Ectosymbiosis A symbiotic state in which one organism attaches to the surface of another.

Ectotropic Virus Virus that can only replicate in host cells.

Edaphic Factors The physical, chemical and biological characteristic of the soil that together form an important component of the habitat because of their influence on plant distribution. The main edaphic factors are water content, pH, organic matter and soil texture.

Edema An increase in the volume of interstitial fluid due to a disturbance in the mechanisms involved in fluid exchange.

Edition Enzyme Enzyme responsible for Proof-reading the DNA and correcting errors in replication of following from mutation.

Edman Chemistry A sequence of chemical reactions generally used as the basis for spinning cup, solid phase, gas phase and micro-sequencing of proteins. Each complete cycle of reactions result in the removal of one amino acid shorter.

Effective Population The number of reproducing individuals in a population.

Effector Cells A functional concept which is context means those lymphocytes or phagocytes which produce the end effect.

Effector T Cells T cells that have an effect on specific antigens (such as cytotxic cells) as opposed to regulatory T cells that have regulatory effects on other lymphocytes.

Effector Any organ, cell, or organelle that moves the organism through the environment or else alters the environment to the organism's advantage. Examples include muscle, bone, and a wide variety of exocrine glands.

Egg A gamets specialized for the storage of food, as opposed to a Sperm which is specialized for mobility and which fertilizes the egg. The female gamete.

Ehiers-danios Syndrome Disease in the human resulting from an autosomal dominant gene (although X-linked and recessive genes are known). It is Characterised by hyper elastic skin, lax joints and poor wound healing.

Elastase Sterine protease which will digest elastin and ollagen type4; inhibited by A_1-protease inhibitor of plasma.

Electro-bolting The electrophoretic transfer of macromolecules (DNA, RNA or protein) from a gel in which they have been separated to a support matrix such as a nitrocellulose sheet. An alternative to the capillary transfer usually used in techniques such as Southern and Northern blotting.

Electrochemical Gradient Gradient that gives the plasma membrane its electrical potential. Both the concentration gradient of the ions and the membrane potential of the membrane may contribute to this gradient.

Electrodialysis A technique used for the deionization of solutions. The sample is placed in a compartment limited by ion-specific semipermeable membranes, across which a potential is applied. Charged ions move to the relevant electrode, whereas larger molecules or uncharged compounds are retained within the chamber. Electrodialysis is used on a large scale for descalination of sea water. It I also used for deionizaton of why permeates, cell hydrolyzats, etc.

Electroinjection A method for direct introducing of foreign genes into intact higher plant cells suing electric field pulses.

Electron Donor Molecule, such as that within the electron transport chain of mitochondrial energy conversion, which provides electrons.

Electron Microscopy A technique for visualizing material that uses beams of electrons instead of light rays and that permits greater magnification than is possible with an optical microscope. Resolutions of –10 A° are attainable with biological materials.

Electronegative Atom An atom with a tendency to gain electrons.

Electrophoretic Transformation Attempts to introduce DNA by electropheresis into germinating seeds of barley and to cell and seeds of various other species with a suitable embryo have so far proved unsuccessful. However, as yet, it is still not possible to decide just how valuable electrophoresis is as a technique.

Electrophoretic Trnsport Through Membranes Movement of charged particles in solution through membranes such as the plasma membrane as result of the existing electrical potential across the membrane.

Electroporation A process uses a precisely controlled electric pulse to induce DNA uptake by protoplast or tissue. DNA enters the cell through pores that are thought to be created in the cell's surface in response to the electric field. During electroration, and initial potential "V" is applied to a cell suspension placed between two electrodes separated by a distance "d". This generates and "optimal" or "critical" electrical field (E= V/d) that is applied for an optimal time.

Electrotonic Potential Inneutrons, a hyperpolarization or small depolarisation induced by the application of a small electrical current.

Elisa (Enzyme-linked Immunosorbent Assay) Sensitive technique used to detect small quantities of specific proteins by antigenic means. A clorimetric reaction, often based on a peroxidase-conjugated antibody used to bind the protein, provides easy quantita of the result.

Ellipsosome Membrane-bonded compartment containing cytochrome-like pigment and found in the retinal cones of some fish.

Elongation The formation of the peptide chain during which a ribosome moves along a molecule of mRNA from the 5'- end to the 3'-end. A specific aminoacyl-tRNA binds to each codon, as it comes into contact with the ribosome, in a position that places the amino acid it carries at the functional centre where the peptide bonds is formed.

Embedding The sealing of tissue prepared for permanent microscope slides in a solid block of paraffin wax prior to sectioning. After cleaning, tissue are placed in two or three baths of molten paraffin wax. When the tissue is completely in filtered by the wax it is allowed to barden. As the wax is opaque the block must be marked to ensure correct orientation when sectioning. Tissue prepared for electron microscopy may be embedded in araldite and cut with a diamond knife.

Embryo Culture (1) Denotes a culture in which the explant was n empbroyo. Embryo cultures have been used to obtain viable off-spring from seeds with a tendency for embroyo abortion or when viable seed are limited in number (2) cultures in which embryos are induced to form [embryogenesis] whether in suspension or on a variety of explants or cultures on solidified media. Ore correctly, these nonzygoti or somatic embryos are termed embryoids.

Embryo Transfer It is a method by which embryos are collected from a donor females and transferred to a recipient female which serves as a surrounding mother for the remainder of pregnancy. Such techniques have been applied to nearly every species of domestic animal, to many species of wild life and exotic animals, as well as to human and other primates. The technique are non-surgical procedures for obtaining the embryo at and early stage of development, cryopreservation and micromanipulation of the developing embryo.

Embryo-Cell One of two cells formed from first division of fertilized egg in certain plants, developing later into embryo, the other developing into suspensor.

Embryo-sac Mother-Cell A cell of hypodermal origin which undergoes a tetrad division to form four cells one of which develops into the embryo-sac or megaspore.

Embryonic Induction Determination of the development fate of a particular area of an embryo by a process at least partially external to the cells themselves embryonic cells that are induced must be competent to respond to the inducer.

Enamel The hard white outer coating of the teeth of vertebrates. It is a protective layer, epidermal in origin.

End Plate A flettened nerve ending that occurs at the junction of a motor axon and a muscle cell. It transmits nerve impulses in a way similar to that of other synapses.

End Product A chemical compound that is the final product of a sequence of metabolic reactions.

End-Plate Potential (EPP) A brief localized depolarization of potential change across the membrane in the motor end-plate region of a muscle fibre, at a neuromuscular junction, a neurotransmitter substance is released from the presynaptic nerve ending a on stimulation by an impulse an increase the permeability of the postsynaptic (muscle) membrane to ions, causing the EPP.

Endergonic Reaction A chemical reaction with a positive standard free energy changes; an energy-consuming reaction.

Endoderm (Entoderm) The innermost germ layer of most metazoans (including vertebrates) that develops into the gut lining and its derivatives. It also forms the yolk sac and allantois in birds and mammals.

Endogamy The fusion of gametes produced by closely related organism.

Endogenous Inhibitor A substance produced by an organism capable of inhibiting its own growth or some specific physiological function or that of another organis exposed to it even at low concentration.

Endoglucanase An enzyme that cleaves a plysaccharide such as cellulose at a point within the macromolecular substrate, rather than at the ends.

Endomixis Replacement of the macronucleus by a product of the micronuceus, in infusoria. Now believed to occur only at Autogamy.

Endoplasmic Reticulum A system of membrane-bounded tubes and flattened sacs, often continuous with the nuclear envelope, found in the cytoplasim of eukaryotes, Exists as rough e. r., studded with ribosomes, and smooth e. r., lacking ribosomes.

Endopolyploidy The occurrence of cell nuclei in an individual representing different degrees of polypolidy. The term maybe more concisely written as Endoploidy.

Endoreduplication Chromosome duplication without breakdown of the nuclear envelops, as occurs in the formation of some polypolid nuclei.

Endorphins Group of large pllypeptides that have been isolated from the pituitary gland and believed to have a morphine-like activity in binding to the surface of nerve cells in the brain and reducing painsensation or altering mood.

Endoscopic Describing the type of development of a plant embryoin which the inner cell formed by the first division of the zygote develops into the embryo, while the outer cell develops into the suspensor. It is seen in many pteridophytes and in all seed plants Compare Exoscopic.

Endosymbiont Theory Theoretical evolution of mitochondria and chloroplast from endoymbiotic prokaryotes which originally existed within primitive eukaryotic cells.

Endothelium The tissue lining the blood vessels and heart, it consists of a single layer of thin flat cells fitting very close together with only a little cement substance between them. In capillaries it is the only blood and the fluid bathing the cells. Water and all dissolved cells by an amoeboid movement known as diapedesis.

Energy Crop Selected plant species grown and harvested specifically as biomass to be used as a source of energy and converted into a fuel or bulk chemical. The major energy crops include fast-growing or coppiedness (willows, poplars and eucalyptous species). Sugar crops (maize or cassava), inulin crops (chicory or Jerusalem artichoke), aquatic plants including algae and water weeds (water hyacine and emergent species such as rushed and reeds).

Enhancer Element Eukaryotic promoter element that increases the transcriptional efficiency of a nearby gene. The first enhancer element discovered was the 72-base pair tandem repeat of Sv40 DNA. Deletion of this element, which is located upstream of the CAP site of the early viral genes, reduces early gene expression by a factor of atleast 100. The molecular basis of enhancer activity is not understood.

Enolase An enzyme that catalyzes the inter conversion of phosphenol pyruvate and 2-phosphogyceric acid. This reaction is one of the intermediate steps in glycolysis.

Entelechy A hypothetical principle controlling purposive adaptation and evolution.

Enterobacteriaceae A large family of gram negative bacilli that inhibit the large interestine of mammals. Commonest is escherichia coli; most are harmless commensals but other can cause intestinal disease.

Enteroviruses Obsolete term for Picornaviruses a group of single stranded RNA virus of small size. The virions have cubic symmetry. Because many priocornaviruses, such as poio virus, were found in the intestine they were termed enteroviruses.

Entomopox Virus An insect virus which bears great morphological similarity to the poxviruses of vertebrates. They are characterised by spindle-shaped to oval virus particles, surrounded by 2-3 limiting membranes, and occluded in large oval proteinceous crystals formed in the cytoplasm of infected cells.

Enucleation Removal of the corpusluteum.

Envelope (Bacterial) The cellular component that encloses the cytoplasm, including the cytoplasmic membranes, cell wall, and capsule.

Environmental Chamber A controlled environment cabinet (incubator) in which temperature, light quality, intensity and duration; and preferably also the relative humidity and airflow are controlled.

Enzyme Activation A mechanism in which the activity of an enzyme is increased by a direct effect on the enzyme, rather than due to new protein synthesis. It may be brought about through binding of an activator molecule at an allosteric site on the enzyme resulting in a change in the enzyme configuration, which leads to a change in the shape of the active site.

Enzyme Analysis A technique in which a specific catalyzed by an enzyme is used to determine either the amount of enzyme or substrate present in a sample that may be highly complex.

Enzyme Equilibrium An expression which describes the state of an enzyme-catalysed reactions in terms of the rates of the forward and back reactions.

Enzyme Extraction The removal of enzymes from contaminating materials in order to increases their specific activity.

Enzyme Fermentation A process in which a microorganism is grown as a source of an industrial enzyme on a large scale. Industrial enzymes are produced by either solid substanrate cultivations using fungal sources or conventional batch submerged culture techniques for bacterial sources.

Enzyme Inactivation at High Temperature In certain animals, e.g. Himalya rabbits and Siamese cats, melanin is produced at extremities, tips of ears, tails, nose, and on feet here body temperature.

Enzyme Kinetics Rate of a reaction controlled by an enzyme. At low substrate concentrations the reaction rate (V) is always proportional to the substrate concentration ([S[), but when the substrate is present Is present at high concentration enzyme saturation will occur (V_{max}). The Michaelis-Menten equation expresses the rate of a reaction as $V = V_{max}$ (S)/(K_m + [S]), where K_m is the michaels constant and is equal to the concentration of the substrate when the speed of the reaction if half its maximum.

Episome Indicates the independently replicating genetical unit that are often associated with the chromosome or the nuclear material of certain organism. These are foreign in origin and are capable of imparting genetical charges in the host, e.g., temperate phases causing lysogenic conversion.

Epistsis A kind of interaction between genes belonging to different pairs of alleles, the dominant allele in one of the pairs preventing the dominant allele in the other pair from expressing itself. Thus, gene A may be epistatic over gene B.B is then said to be hypostatic to A.

Epistatic T-DNA Interactions The inactivation of bacterial nopaline synthase [nos] and kanamycin resistance genes carried by T-DNA [T'-DNA-1] on introduction of a second T-DNA [T-DNA [T-DNA-II]. T-DNA-II carried part of a nos gene and other sequences homologous to t-DNA-1. The phenotypes of the gnees carried by T-DNA-II are not suppressed, i.e., suppression is not observed to eco-ordinated, lack of expression of T-DNA-I is correlated with methylation of the DNA.

Epithelium A lining of cells, as in the surface of the skin or lining of the gut, usually derived from the embryonic endoderm or ectoderm, but exceptionally derived from mesodersm as with kidney tubules and mesohelium lining body cavities. These cells frequently have a high nuclear to cytoplasmic ratio when compared to fibroblasts.

Epstein-Barr Virus DNA herpes-type virus that causes infectious mononuclease (glandular fever) and is also causatively involved in Burkitt's lymphoma, amalignant tumour of lymph glands common in Central Africa.

Equational Division The second cell division cycle of meiois in which the daughter cells retain the haploid chromosomes number of the parent.

Equilibrium Constant A particular kind of ratio of the concentrations of products to reactant when equilibrium has been reached in a chemical reaction, Designated by the symbol K_{eq}.

Equilibrium Density Centrifugation Separation of cell organelles or macromolecules of different densities by centrifugation in a different densities by centrifugation in a density gradient. The gradient is a solution of sucrose, or cesium chloride, that increases in concentration from top to bottom of the centrifuge tube.

Equilibrium Frequency In a population that gene frequency which varies nondirectionally about a mean by an amount described by the standard deviation under conditions of unchanging selection pressure and mutation rate, with no inter mixing from other populations.

Eragastoplasm Now obsolute term for the Endoplasmic reticulum Groups of ribosome, no w known as polysomes, were recognised as being a characteristic of this membrane network, and they were then referred to as ergosomes.

Ergies Organic biocatalyzers including enzymes, ergones, vitamins and hormones.

Erlenmeyer Flask A wide-necked variety of conical, flat-bottomed flask commonly used for medium preparation. Smaller versions are used as culture containers. Named after E. Erlenmeyer (1825-1909).

Error-checking Protocol In the transmission of data in potentially nosy media it is important to check that the data received corresponds with the data that was sent. The effects of noise or corruption can thereby be detected ad the data resent. A variety of methods are available that achieve this and are incorporated at many levels in a computer system (e.g. reading/writing to magnetic disk or tape, transmission of data over communications lines).

Error-prone Repair Type of DNA repair in which DNA chain growth occurs across damaged DNA segments, but some of these new strands are themselves defective. Such error-prone repair appears to be the chief source of genetic mutation in the bacteria in which it occurs.

Erythrocyte (red blood cells) The biconclave, nonnucleated and agranular mature cell of vertebrate blood that transports oxygen to the tissues in combination with hemoglobin, the pigment responsible for the red color of fresh blood.

Erythrogenic Tozin Toxin produced by strains of Streptococcus pyrogenes responsible for scarlet fever.

Erythroleukaemix Cell Abnormal precursor of erythrocyes which can be grown in culture and induced to differentiate by treatment with, for example, DMSO.

Erythropoletin Glycoprotein hormone (M-460000) and produced in response to reduced oxyen tension, and its production is enhanced by blood iron (which reduces oxygen-carrying capacity) and is depressed by erythrocyte excess.

Established Culture (1) Achievement of an aseptic viable explant, synonymous with Stage I culture [culture establishment] or transplant, synonymous with Stage IV culture [establishment in soil]. (2) A suspension culture subjected to several passages, adjusted so that cell number per unit time is constant from subculture to subculture.

Estrogen Collective term for the female sex hormones (natural; estradiol, estrne, estriol; synthetic; ethynylest radio land stilbestrol) responsible for the development of the accessory sex structures and the secondary sex characteristic in the female. Estrogenic hormones also affect calcium and phosphorus metabolism, bone and lipid metabolism.

Estrous Cycle The repeated series of changes in reproductive physiology and behaviour in female mammals (other than some primates), culminating in estrus.

Etheogenesis, Ethiogenesis Male parthenogenesis; the formation of an organism from a male gamete without fertilisation having taken place.

Ethnic Group A population group united by manners and customs and an associated with a particular geographic region.

Ethnobiology The study of the relation between man and his surrounding animal life. Ethology the study of whole patterns of animal behaviour in natural environments, stressing the analysis of adaptation and evolution of the patterns.

Ethylene Oxide A colorless gas having remarkable penetrating power that is frequently used a a sterilizing or decontaminating agent. It is effective at room temperature, although a certain level of moisture is required for optimum effectiveness. The flammability hazard is greatly reduced by diluting the gas with carbon dioxide or fluorocarbons such as Freon.

Eubacteria One of the two major groups of bacteria. Eubacteria have fundamental features that are considered to be typical of most bacteria.

Eucaryote An organism characterized by a cellular organization that includes a nuclear membrane and other internal membrane organelles such as mitochondria and mitotic apparatus.

Euchromocentre The heterochromatin on either side of the centric construction.

Eugenic Progosis Anestimate of the probability of a given disease or defect being inherited.

Euhermaphrdite A normal hermaphrodite plant in which both anthers and gynoecia are well developed in all flowers.

Eukaryotes Organisms, including animals, plants fungi, and some algae, that have (1) chromosomes enclosed within a membrane-bounded nucleus and (2) functional organelles, such as mitochondria and chloroplasts, in the cytoplasm of their cells.

Eukaryotic Cells Cells in which the genetic material is contained within a distinct nucleus. This includes all organisms and cells except bacteria and blue-green algae.

Eutrophic Describing takes or ponds that are rich in nutrients and consequently are able to support a dense population of plankton and littoral vegetation.

Evaporation The removal o a liquid by conversion of the vapour.

Evaporator An apparatus designed as a means of removing liquid; in biological systems the liquid is usually water.

Evapotranspiration Evaporation from soil surfaces, lakes, and streams, and by transpiration from plants into the atmosphere.

Evocation Induction of cellular differentiation or a pattern of development within an embryo by an evocator, deemed to be a molecular compound produced by the organizing centre relevant to the responding tissue. Evocators are more or less equivalent to the hypothestical substances more often referred to as morphogens or inductors

Excitation-contraction Coupling A process by which excitation of the muscle fibre membrane at neuromuscular junction results in contraction of a muscle.

Exoenzymes An enzyme excreted by a microorganism into the environment Also called Extracellular Enzymes.

Exogenote That portion of the bacterial donor chromosome which is transferred during conjugation.

Exogenoous DNA DNA that has been derived from a source organism and has been cloned into a vector and introduced into a host cell. Also reterred to as foreign or heterologous DNA.

Explant Culture A technique for transforming plant cells by which tissue vector engineered to contain foreign DNA.

Explant Donor The source plant or mother plant from which the explant used to initiate the culture is taken.

Etramedial Hybridity Quotient A measurement of the extent to which measurement of the extent to which the hybrid departs from the means between the two parents.

Extrinsic Isolating Mechanism Any barrier in the environment, such as a river bed, desert, or ocean, that isolates one population from another and thus prevents mating between them.

Exudate Fluid, often from formed elements of the blood, discharged from tissues to a surface or cavity.

F

F factor The fertility factor in the bacterium Escherichia coli, it is composed of DNA and must be present for a cell to function as a "male" or donor in conjugation.

F_1 (1) Abbreviation for the first filial generation, usually the hybrid between two homozygous types (2) The progeny in the second generation of a of cross. In plants, F_2 progeny are obtained by sowing seeds through self-polinating the F_1 flowers. In animals the progeny is raised through brother-sister (sib-) mating.

Fab Fragment Part of an immunoglobin molecule containing the antigenbinding site (Fragment antigent binding) and produced by treatment with the enzyme papain. It is composed of one light chain and part of one Heavy chain by a disulphide bond.

Facilitated Diffusion A passive transport of molecules across a cell membrane along a concentration gradient, mediated by carrier molecule or complexes. No energy is expanded in this process.

Factorial Experiment One in which all the treatments of agents under investigation are varied simultaneously, and combined in such a way that any derived effect of one or a group of them may be isolated and separately evaluated.

Facultative Pathogen A microorganism which is capable of growth and reproduction in either a nonliving medium or living host; in the latter instance, adisease condition of the host may arise; of Obligate Pathogen.

Fallopian Tube The first portion of the Mullerian duct in mammals which conveys ova from the peritoneal cavity to the utrus. The upper portion of the oviduct.

Familial Hypercholestrolaemia (FHC) Human hereditary disease in which Lowdensity Liporotein is at an elevated level in the blood plasma. It is inherited as an autosomal dominant and leads to atherosoclerosis.

Family A group of individuals within a breed, all of which have pedigrees which trace directly in the female line to a common ancestress called the foundress of the family. At times used in reference to the male line of descent, in which case it is used interchangeably with line of breeding.

Farina An obsolete synonym for pollen.

Fastidious Organism An organism that is difficult to isolate or cultivate on ordinary culture media because of its need for special nutritional factors.

Fatty Acid Oxidation Process by which acetyl coenzyme. A oxidizes fatty acids in the mitochondria of animals and plants cells. Through this metabolic pathway neutral fat is utilized to release energy for the cell.

Fatty Acid A straight chain of carbon atoms with a COOH at one end in which most of the carbons are attached to hydrogen atoms.

Feedback (End-product) Inhibition Inhibition of the enzymatic activity of he first enzyme in a metabolic pathway by end product of the pathway.

Feeder Cell Cell that when added to another cell population in tissue culture facilitates the growth of the latter frequently the feeder cells have been previously irradiated to prevent their division in the culture.

Feline Leukaemia Virus RNA virus associated with leukaemia in cats.

Female Plant A pistilate plant in a dioecies species. It produces the female gametes, the egg.

Female Sterility A type which produces no viable eggs, but ample sperm in the pollen; due to complementary genes.

Feredoxins A group of red-brown proteins of green plants, blue-green algae, and some bateria. They contain non-haem iron in association with sulphure at the active site. They are strong reducing agents (very negative redox potentials) and function as electron carriers, and function as electron carriers, for example in photosynthesis and nitrogen fixation.

Fermented Dairy Products A wide range of processed foods prepared by growing microorganisms on whole or separated and/or reconstituted milk. Product include acidophilous milk, buttermilk, cream cheese, yoghurt, etc.

Fermenter Agitation A method of mixing or stirring the contents of a fermenter in order to prevent, to facilitate rapid gas exchange (oxygen uptake in particular), and to establish uniform temperature and nutrient distribution.

Fermenter Control Advanced fermentation process maybe equipped with a wide variety of automatic control and/or moitoring facilities using microprocessors. These includes: (1) control of temperature, pressure and pH by closed loop feedback; (2) control of foaming b on/off addition of an antifoam; (3) automatic sequencing to ensure correct adherence to protocols for sterelization, cleaning and filling, as well as harvest operations; (4) control of dissolved oxygen concentrations; (5) automatic feeding of nutrients in response to dissolved oxygen concentration or biomass level, or to feed nutrients automatically according to a preset schedule; (6) control of the respiratory quotient in order to optimize cell mass production.

Fermentograph An automated device used to measure the production of carbon dioxide in order to assess the activity of baker's yeast in a sample of dough.

Ferric Ethylenediamine Tetraacetate, Sodium Salt [NaFe. EDTA $C_{10}H_{12}FeN_2NaO_8$ mw 367.07]: A microelement iron and sodium salt added to some plant tissue culture media. It may be substituted for $FeSO_4 + Na_2$.EDTA however, if substituted on an equimolar basis, while the same amount of Fe will still be available, half the Na will be lacking and may have to be added in the form of another sa;t/

Ferritin Iron-containing compound that is used as a labeling molecule in electron microsocopy due to its high electrons cattering properties (following from its high iron content). Ferritin is either used directly to study cell surfaces or more commonly. It is conjugated to antibodies to form an electrodense probe for specific antigen molecules. Structurally ferritin consists of a protein (apoferritin) combined with colloidal ferric hydroxide-ferric phosphate micelles.

Fertility Restorers Restorers have gene that restores fertility to a cytoplasmic male sterile line.

Fertilization Membrane A membrane which grows outward from the point of contact of egg and sperm and which rapidly covers the whole egg surface.

Fertilization The union of a male gamete (sperm) with an egg (female gamete) to form an embryo.

Fertilizin A soluble substance produced by the egg and assumed to play and essential part in fertilization as a chemical link between egg and sperm.

Fertilization Membrane A membrane which grows outward from the point of contact of egg and sperm and which rapidly covers the whole egg surface.

Fertilization The union of a male gamete 9sperm) with an egg (female gamete) to form an embryo.

Fertilizin A soluble substance produced by the egg and assumed to play and essential part in fertilization as a chemical link between egg and sperm.

Fertilization Membrane A membrane which grows outward from the point of contact of egg and sperm and which rapidly covers the whole egg surface.

Feulgen Test It is a chemical test for DNA. The test is made by first hydrolyzing the chromosomes with the Schiffs reagent and forms characteristic colouration.

Feulgen-positive Characteristic of DNA and DNA-containing structures, indicating their positive response to staining with the feulgen

Feulgen-stain Important cytochemical test for DNA. It utilizes Schiff's reagent and gives a purplish pink colour when positive.

Fibrin An insoluble protein material that adds blood clotting. It is not present as such in any quantity in blood but is formed from fibrinogen, which is normally present in blood plasma. The conversion from soluble fibrinogen to insoluble fibrin is brought about by the enzyme thrombin. If fresh blood is rapidly whipped a stringy mss of fibrin is obtained.

Fibroblast A cell that produces fibres in connective tissue. Usually they are long flat cells found alongside the fibres.

Fibroblastic A proliferating precursor cell of the mature differentiated fibrocyte. These cells are usually spindle-shaped (bipolar) or stellate (multipolar) and are arranged in parallel arrays at confluence if contact inhibited. Cells are usually migratory with processes exceeding the nuclear diameter by threefold or more.

Fibroin Protein that is the major component of silk, a material used by spiders to build webs, and by insects to construct larval nests and cocoons.

Fibroplast A cell that produces fibres in connective tissue. Usually they are long flat cells found alongside the fibres.

Fibrous Proteins Insoluble proteins that serves in a protective or structural role, and in which the protective or structural role, and in which the polypeptide chain is extended or coiled along one dimension.

Ficoll Synthetic sucrose polymer used widely in separation techniques and centrifugation in cellular biochemistry.

Field Test An evaluative test whereby the field performance of experimental plants is assisted in comparison to controls.

Filamentous Characterized by thread like structures.

Filamin Long Flexible Fibrous protein that has a tendency to form dimmer in solution. It is important in its role of forming cross-link between neighbouring action molecules, thus changing viscous fluid containing only action molecule in solution, into a solid gel with the addition of filamin another protein with a similar action to filamin is called A-action.

Filter Hybridization Exposure of DNA bound as denatured single strands to an immobilizing filter material such as nitrocellulose paper, to a solution containing radioactively labeled probe DNA or RNA-Subsequently washing leaves only base-paired double strained molecules and so identifies complementary sequences in the original sample.

Filter, Bacterial A special type of filter through which bacterial cells cannot pass.

Filtration Passage of a liquid through a porous membrane for the removal of particles. Also the use of certain types f glass to remove particular wavelengths of light.

Fine-structure Genetic Mapping Analysis of recombination within a gene sequence down to the level of the single Nucleotide. Recombination occurs within gene sequences as well as between gene sequences, and the location of particular base changes in mutant alleles can be determined by this means.

Fingerprinting Technique widely used in the determination of amino acid sequence between proteins that are clearly related (e.g. products of mutants alleles). The technique involves the enzymic breakdown of the protein into peptides and separation of such peptides by electrophoresis or chromatography often two-dimensionally, permitting a comparison of the pattern obtained after migration of the peptide.

Fission Yeast Yeast that replicates by fission with the formation of a cell plate in the middle of the rod-shaped cell. This is in contrast to propagation by bud formation in Budding yeast. The best known fission yeast, widely used in studies on the cell cycle, is Schizosacharomyces pombe.

Fission, Binary Single nuclear division followed by division of the cytoplasm to form two daughter cells of equal size.

Fistulinacease Basidiocarp: Annual, pileate with a lateral stipe, fleshy and moist; the hymenophore consists of a layer of closely packed parallel tubeules, Fistulina hepatica (the edible 'beef-steak fungus') causes Bron Oak. It forms a fan-shaped ortongue-shaped basidiocarp whose upper surface is rusty brown, and whose lower (hymenial) surface is pale yellow when unbroken; the reddish monomitic context of the pileus exudes a red liquid when cut.

Fitness (1) The number of offspring left by an individual as compared to individual of different genotypes. (2) A relative quality the difference in the reproduction and survival of one organism in a given environment over another organism: Fitness = 1 – selection cofficient W = t – s

Fixation of Alleles State in which all members of a population are either Homozygous or hemizygous for a particular allele, all other alleles at that locus having been lost in the course of evolution. Flagellum a long whiplike organelle used in the movement of certain organisms and cells, it is structurally similar to a cilium.

Flame Bulb A cub-shaped mass of protoplasm bearing a tuft of cilia projecting into the cavity of the cup, found at the closed inner end of a protonephridium.

Flame Sterelize To sterelize tools or instrument usually by heating them in a flame until they glow. Thus procedure is used in conjunction ethanol (70%) emersion or ethanol (95%) dip. The most common means of sterilizing forceps and scalpels used in sterile tissue culture manipulations.

Flanking Sequence Sequence that is either upstream or downstream, on the same DNA molecule, from a particular definitive sequence such as a gene. Enhancer sequences frequently flank gene-coding sequences.

Flareback (pest resurgence) The (sometimes dramatic) rapid increase of a target insect pest population subjected to an insecticidal treatment which also destroys its associated, and possible regulative natural enemies.

Flavonoid One of a common group of plan compounds having the C_6-C_3-C_6 chemical skeleton in which C_6 is a benzene ring.

Flavin Nucleotides Nuleotides coenzymes FFMN and FAD) containing riboflavin.

Floculent Masses of cells floating, or settled to the bottom, in a liquid medium.

Flora All of the plants found in a given area.

Flora Microorganism present in a given situation, e.g. intestinal flora, normal flora of soil. Also called Biota.

Floricane The flowering and fruiting stem (2nd year) of biennialy producing plants such as Rubus species. The first year canes are vegetative and are called primocanes.

Florigen (flowering Hormone) A hypothetical plant hormone that has been portulated to account for the transfer of photoperiod stimulates from the leaves to the apex here flowering have so far been un successful. Florigen. A plant hormone (not yet isolated) involved in the conversion of a vegetative shoot apex to a flower.

Flower The total reproductive structure of an angiosperm; its basic parts include the calyx, corolla, stamens, and carpels.

Fluctuation Test Generally test originally devised by Luria and Delbruck in 1943. The test was devised to determine whether mutation in bacteria occurred trandomly and non-adaptively, or in response to specific environment demand. In the test these worker compared the number of a favourable mutation occurring in many small, parallel cultures of a bacterium with the numbers occurring in one large culture.

Fluctuation Test Test to determine the development of mutants in a culture. The test is based on the concept that matuation is a completely random event, and that spontaneous mutation would result in a large fluctuation in the number of mutants in a series of identical cultures.

Fluid Drilling A mechanical procedure for planting seed; pre-germinated seeds are suspended in a gel and sowed through a fluid drill seeder. This technology is potentially adaptive for sowing artificial seed (somatic embryo or embryoidsa.)

Fluorescein isothiocyanate A compound used to stain plant protoplasts; as in the identification of fusion products.

Fluorescein-conjugated Antibody Anti body sample that has been conjugated to a flurescent dye and maybe used to label cells which specifically bind the antibody.

Fluorescein A fluorescing pigment produced by many pseudomonads and observed by illumination with ultraviolet light.

Fluorescence Microscopy The staining of micro-organism with a fluorescent dye for microscopic study, using ultraviolet light as the illumination source.

Fluorescene-activated Cell Sorter (FACS) Instrument that is able to separate cells that fluoresce from those that do not as they pass down through a narrow or face for scanning.

Fluorescence Light emitted by a substance after absorption of radiation of a shorter wavelength from another source.

Fluorescent Antibody An antibody conjugated with a fluorescent dye that fluoresces on ultra violent irradiation, used in an immunoidentification procedure; may be visualized by UV or fluorescent microscopy.

Flurescence microscopy Microscopy in which the microorganisms are stained with a flurescent de and observed by illumination with ultra violet light.

Flurescent antibody Technique Detection of specific antigens in cell by staining with a specific antibody conjugated with a flurescent dye.

Foetalization The prolongation into post-natal or adult life of foetal characters of remote ancestors.

Fold Back DNA (Snap back DNA) Short repeated sequences of the eukaryotic genome which can fold back on themselves to form duplexes following Denaturation. These sequences can be either dispersed or found in clusters.

Fold-Back The structures formed when a double-stranded DNA molecule containing an inverted repeat sequence is denatured and then allowed to reaneal at low DNA concentrations. The repeated sequences permits the formation of a double-stranded region within each of the separated strands of the original molecules.

Follicle Stimulating Hormone A gonadotropic hormone produced by the anterior pituitary.

Follicle-stimulating Hormone (FSH) A gonadotrophin produced by the anterior pituitary gland. It acts on the ovary to stimulate the growth and maturation of the issue forming folicles and ova, which under the action of luteiniing hormone, mature and are released from the ovary. It is also stimulating in spermatogenesis in males. It has been used, the treatment of female sterility.

Follicular Hormone A female sex hormone formed by the ovarian follicles.

Fomites Inanimate objects or materials that act as intermediate carriers of microbial contamination. Contaminated tools used to assemble a sterile spacecraft could be fomites.

Food Chain A portion of a food web, most commonly a simple sequence of prey species and the predators that consume them.

Food Chain A sequence of organisms by which food energy is transferred from autotrophs to heterotrops each organism consumes the preceding one and in turn is eaten by the following member of the sequence.

Food subweb A group of organism united as a result of being food for the same or similar sets of predators.

Food Web The complete set of food links between species in a community; a diagram indicating which ones are the eates and which are consumed. Forespore A structure than develops in spore-forming bacteria after the active growth period. It later becomes refractile, forms a thick wall, and becomes the resistant endospore.

Fossil Any recognizable structure originating from an organism, or any impression from such a structure, that has been preserved from prehistoric times.

Founder Effect Genetic drift resulting from establishment of a population by a very small number of individuals.

Fraction Sterilization Tyndallization sterilization of culture medium by exposure to steam for 30 minutes each day for three successive days to kill vegetative cells.

Frame-shift Mutation A mutation resulting from the addition or deletion of a single base pair in the DNA sequence of a gene. As a result of this, mRNA transcribed from such a gene is translated normally until the ribosome reaches the point at which the mutation has occurred. From that point on, codons, are read out of proper register and the amino acid sequence bears no resemblance to the normal sequence.

Freeze-dry or Lyophiize To dry in a frozen state under vacuum; as tissues are freeze-dried to obtain a dry weight or to preserve them for analysis.

Freezedrying A technique in which microorganisms are frozen rapidly and then dehydrated in high vacuum directly from the frozen state.

Frequency-depended Selection Selection that is more intense when the individuals undergoing selection are present in greater abundance.

Freund's Adjuvant An emulsion of aqueous antigen in oil; complete freund's adjuvant also contains killed Mycobacterium tuberculosis, while incomplete Frund's adjuvant does not.

Fruit In angioserms, a ripened and mature ovary(or group of ovaries) containing the seeds. Sometimes applied to preproductive structures of other groups of plants, and includes any adjacent parts which may be fused with the reproductive structures.

Fruiting Body A specialized structure in some bacteria (Myxobactria) and some fungi that produce spores.

Fungi Eumycetes; eucryotic cells, frequently filamentous, which lack both a vascular system and chlorophyll; includes molds any yeasts.

Fungus A member of the kingdom Fungi, a (usually) multicellular eukaryote with absorptive nutrition.

G

G Phase A phase of the cell cycle in G1, following mitosis, only one chromatid is present, hence the cells have only half the normal mount of DNA. In g2 duplication is complete but mitosis has not yet occurred.

G-Antigen A particular kind of particles produced during multiplication of certain animal viruses; act as precursor to the complete virus particles. Consist of viral nucleic acid, haemagglutinin and lipids.

Gag-protein The protein of the nucleocapsid hell around the RNA of a retrovirus.

Galactose A sugar found in lactose and many polysaccharides. It is an aldohexose in the blood.

Galactosyltransferase Any one of a group of enzyme that helps to make mature complex oligosaccharides from more simple sugars are finally attached to proteins, and much of this assembly occurs in vesicles of the golgi apparatus. Indeed these enzymes are used as specific markers for the Golgi vesicles.

Gall Gross enlargement and deformity, usually globose, of plant tissue; caused by Agrobacterium tumefaciens.

Gamete Cell which is specialized for fertilization and does not usually develop without it.

Gametic Lethal Factor A gene which renders inviable a gamete which carries it.

Gametic Mutation A mutation occurring during gamete formation.

Gametic Reduction Meiosis which takes place immediately prior to fertilization.

Gametic Tissue Having n number of cronmosomes (Haploid) in contrast to zygotic tissue with 2n (diploid). Some microorganism for the most part are haploid.

Gametogenesis The formation of gametes including meiosis where meiosis immediately recedes the formation of gametes.

Gametophyte The sexual or reproductive generation in plants. This generation has a halved chromosome number, thus representing the haplophase in contrast to the sporophyte, which arises by fertilization and represents the diplophase.

Gametophyte Generation The haploid [n], gamete-producing phase of the life cycle in plants with alternating generations. Gametophytic differences Differences observable in the gametophyte, e.g., waxy vs. starchy pollen in corn and in some other cereals. Most, if not all, of the differences in lower forms (haploid) are gametophytic.

Gama Globulin A fraction of globulin (a plasma protein); identified by electophoresis; contains the majority of immuniglobulins (antibodies); also signifies the antibody containinng fraction of plasma that is used for prophylaxisis therapy.

Gamma-Aminobutyric Acid An amino acid that acts as an inhibitor of neurotransmission. It is founding nervous tissue and in the brain in particular.

Gamma-Counter A device used to measure the amountof Germanys emitted by a sample

Gamma-globin Type of globin present, along with alpha-globin, in some foetal mammals, including the human. It is replaced by beta-globin in the adult.

Gamma-radiation Electromagnetic radiation of short wavelength that is emitted during the radioactive decay of some atomic nuclei. Radioactive sources such as cobalt-60 are powerful emitters of gamma-radiation and as such may be used to induce celluar mutations or to kill cells.

Gamma-richhess Addition of species on a broad geographical scale.

Ganglin A collection of nerve cell bodies, usually bound by a sheath or capsule, In vertebrate the ganglia are located chiefly outside the central nervous system; in invertebrates ganglia occur along the major nerve cords and are the centres of nervous integration.

Ganodermataceae Basidiocarp: annual or perennial, typically fan-shaped and horizontal, eithernon-stipitate or borne on a lateral stipe; the context of the pileus is trimitic, and the underside of the pileus bears a porous hymenophore. Bsidiospores: ovoid, typically flattened at one pole; ornamentations on the inner, brown layer of the sporewall penetrate deeply into the outer, hyaline layer of the spore wall. The organisms occur e.g. on felled timber; Ganoderma applanatum causes heart rots in various types of tree, infection occurring via wounds.

Gap Junction Nexus; portions of the plasma membranes of adjacent cells that contain a space between the two membranes that permits cell-to-cell communication.

Gargoylism Two distinct hereditary disease of the human-hunter syndrome and Hurler syndrome. Both are connective tissue disorders involving storage of glycosaminoglycans and result in facial and other deformities.

Gas Glowcounter Instrument used to measure radioactivity in test samples and in which a sensitive volume of gas ins maintained by allowing constant gaseous flow through the instrument.

Gas Sterilizer An autclave that is designed for or has been modified to permit optimal use of a gaseous decontaminant instead of steam for sterilizing materials.

Ga-liquid Chromatography (GLC) A chromatographic method in which the materials to be analysed are vaporized and carried in an inert gas (e.g. nitrogen) through a stationary liquid phase. The method is suitable for analyzing complex fatty-acid mixtures and any other substance that are volatile at reasonably low temperature.

Gastric Juice An agent of digestion in the stomach secreted by gastric glands situated in the thick stomach wall. It contains two main enzymes; pepsin, which breaks protein down into short polypeptide chains, and renning which coagulates caseinogen to form casein. Gastric juice also contain mucus and has an acid pH. The mechanical and chemical stimulation of the stomach lining by food itself causes secretion of gastric juice and of an hormone (gastrin). This hormone circulates in the blood and causes the gastric glands to secrete hydrochloric acid, thus generating the acid pH of the stomach.

Gated Channel Plasma membranes channel of a proteinaceous nature which can be open or closed, depending on reactions with other components. Some-time voltage-gated channels-are sensitive to action potentials and others-the ligand-gated channels-require specific ligand attachment to reactor proteins.

Gaudens-velans Hybrid (Oenthera) Oenothera lamarckiana is a hybrid of the gaudens-velans complexes.

Geitonogamy Self-pollination in which the pollen is derived from another flower on the same plant.

Gel Electrophoresis Nucleic acid molecules can be distinguished by differences in their chain length, and study of gene organization and expression includes nucleic acid hybridication with a chain length measurement.

Gel Filteration A chromatographic procedure for separation of a mixture of molecules on the basis of size, though the capacity of porous polymers to exclude solutes above a certain size.

Gel Training Procedures whereby proteins, nucleic acids or other molecules are visualized in an electrophoretic gel by staining. Stains such as Coomassive blue are frequently used to stain protein in gels, the initial staining procedures being followed by a regulated destining, until only the desired bands are left stained.

Gel A mixture of thick or firm consistency containing liquid trapped in a solid component, as gelling agents such as agar solidify liquid nutrient media. The process gelation.

Gelatin A protein obtained from skin, hair, bones, tendons, etc. used in culture media for the determination of a specific proteolytic activity of microorganisms or for the preparation of a peptone.

Gemma An organ of vegetative reproduction produced in mosses, liverworts, and certain pteridophytes, e.g. Lycopodium selago. They often form in groups in receptacles called gemmacups and eventually become detached from the parent to form new plants.

Gemmation The production of specialized buds which are capable of developing into a new independent organism.

Genasthenia, Theory of In addition to their accepted qualitative properties, this theory attributes to the genes quantitative characteristics, or functional valency, which may be reduced in a heterozygote or in a partly foreign plasma.

Gene Cloning Technique hereby individual gene sequences may be isolated from the genome of an organisminserted into virual or bacterial Plasmid Conning Vectors, and thus produced in large numbers.

Gene Expression Manifestation of a gene or genes in the phenotypic character of an organism. Single genes are said to express themselves in terms of particular genetic traits, which are most frequently a result of the gene determining the amino acid sequence of particular protein such as an enzyme.

Gene Frequency The Frequency of a gene in a population, expressed as a fraction or a decimal fraction, with prepresenting the normal allele and q representing the mutant. In controlled matings of homozygous types, p = q= ½ and p+q=1. Also q=1-p.

Gene Isolation The removal of the genetic information, in the form of a DNA sequence, from a selected organism in order to study its structure or insert it into a vector during gene manipulation.

Gene Locus Positive on a particular chromosomes on which a specific gene isolated, either relative to other genes which are adjacent on the same chromosome or in terms of a particular mapping distance from the centromere of the particular chromosomes.

Gene Magnification Henomenon in the which the number of copies of the ribosomal Genes is increased in mutants of Drosophila. Such mutants are termed bobbed, since they also have reduced number of bristels.

Gene Map Order of nucleotides and genes that is deduced from genetic recombination experiments. Individual genes can be assigned to specific loci within a genome or chromosome a result of gene mapping.

Gene Reduplication The duplication of genes in the chromosomes at cell division so that each daughter gene and chromosome is an exact replica of its original.

Gene Regulatory Protein Protein that functions by altering the transcriptional state of a specific gene. The repressor protein that regulates the activity of the lac operon in E. coli is one such protein, and they are widely believed to be an important aspect of eukaryotic gene regulation also.

Genetic Linkage Mapping This form of mapping is used to construct genetic linkage maps of individual chromosomes and also to localize genes involved in genetic disease.

Gene Starvation Theory A theory put forward by Thompson (1938) suggesting that to reproduce itself normally a gene must have available in its vicinity all the necessary 'parts'. Lack of any 'part' during the stage between successive nuclear division is suggested as the cause of mutation, or alternatively of gene deficiency, at the 'starved' locus. On this theory, differential nutrition in the early stages of segmentation of the fertilized ovum gives rise of cells from which are produced the primordial of the differentiated adult tissues.

Gene Switch Change in a cell or organism from the expression of one type of gene, or set of genes, to another related one, or set. Classic examples f gene switching occur during the developmental processes of higher organism.

Gene Synthesis The production of base sequences in lengths of DNA which can subsequently be expressed as peptides after insertion into a suitable host.

Gene Transfer Technology Genes transfer from one organism to another molecular biologists used vectors. Vectors are the carriers used to pass genes to a new host, and they can mediate the entry, maintenance, and expression of foreign genes in cells. Factors used to transfer genes include virsus. Plasmids, and mobile segments of DNA called transposable elements.

Gene In the genome of an organism, a sequence of nucleotides to which a specific function can be assigned e.g. a nucleotide sequence coding for a polypetide, a nucleotide sequence specifying a tRNA, or a nucleotide sequence required for the proper transcription of another gene.

Generational Sterility Basic, inherent sterility due to such causes as an unbalanced chromosome number, or lack o homology as between the two chromosomes sets in an interspecific hybrid, as opposed to sterility due to malformed anthers or pollen incompatibility.

Generative Nucleus The gemetic nucleus of a pollen grain as distinct from the vegetative or tube-nucleus.

Genetic Assimilation Process by which a phenotype character becomes genertically fixed in a population.

Genetic Association A term used to describe the condition where particular genotypes are associated with other phenomena, such as particular diseases.

Genetic Complementation The condition wherein two recessive mutants, defective in different ways for the same cisern (and hence, the same polypeptide) produce in combination(in mixed infections, heterogenotes, heterokaryons or heterozygotes) a trans dihybrid which only approaches the wild phenotype (pseudowid) and which is therefore, different from the corresponding cis double heterozygote.

Genetic Deficiency The result of a chromosomal aberration in which a piece of a chromosome is lost.

Genetic Drift A phenomenon in which there is chance fixation or loss of certain mutations in a population of a small size. The genetic drift accounts for the preponderance of certain areas of the world.

Genetic Engineering or Recombinant DNA Technology Technology involving man-made changes in the genetic constitution of cells [apart from selective breeding]. This technology usually employs a vector {such as he Ti plasmid of Agrobacterium tumefaciens] for transferring useful genetic information from a donor organism into a cell or organism that does not possess it. Genetic engineering has many potential uses.

Genetic Engineering Genetic engineering is the most fundamental mechanics of biotechnologies and is a recent off-shoot of biotechnological research. It technology. The technique overall involves into two steps: (1) the in vitro incorporation of the gene or segment of DNA of interest into a small, self-replicating chromosomes and, (2) the introduction of the recombinant minichromosome into a host cell where it will replicate. Step one involves synhesis of recombinant DNA and step two is he Gene cloning.

Genetic Equilibrium The condition of a population in which successive generations consist of the same genotypes with the same frequencies, in respect of particular genes or arrangement of genes.

Genetic Homeostasis Process by which a population of organism persists sudden gentic changes by the equilibrium of its genetic composition and gene pool.

Genetic Industrion Stimulation of a specific gene by an inducer molecle. In bacterial operons more than one gene may be induced simultaneously by the same inducer. Inducers may act directly or indirectly, usually through an effect on the RNA Polymerase or on the promoter region upstream from the gene-coding sequence.

Genetic Information The hereditary information contained in a sequence of nucleotide bases in chromosomal DNA or RNA.

Genetic Linkage The term refers to the fact that certain genes tend to be inheretid together, because they are on the same chromosome.

Genetic Modification A process that results in a change in the genetic make-up of a population.

Genetically Modified (GM) Crops The development in the direction of plant cell and tissue culture research occurred after the discovery of restriction endounuclease enzymes in the early 1970s. These enzymes are responsible to cleave DNA molecules at predicatable sites. They help in removing specific genes of desire which can be modified, an inserted into other DNA strands. The development of watershed of technology has provided the basic tools needed for genetic engineering. The availability of totipotent plant cells that, conceivable, could be altered by insertion of specific genes caused a revolution in plant research. The potential commercial value of such genetically modified (or GM) plants attracted an unprecedented amount of industrial and investor interest in the late 1970s and 1980s which have developed many new plant biotechnology-based companies around the world.

Genetic Polymorphism The continued occurrence in a population of two or more discontinuous genetic variants in frequencies that cannot be a accounted for by recurrent mutation.

Genetic Restriction The term used to describe the observation that lymphocytes and antigen presenting cells cooperate most effectively when they share particular haplotypes.

Genetic Selection Selection of genes cells clones, etc. by man within populations or between populations or between populations or species. The usually purpose is to alter a specific phenotypic character. Such selection usually results in differential success rates of the various genotype relecting many variables including selection pressure and genetic variability in populations.

Genic Explanation of Hybrid Vigour Hybrid vigor is widely explained as the result of a dominance of linked genes. Many dominant genes are involved Different in breads contribute different dominant genes so that the F_1 has more the an either parent.

Genome Analysis Clarification of the homology between more or less related species. This analysis is carried out by crossing and study of the chromosme pairing at meiosis in the F_1 hybrids.

Genome A set of chromosomes herited as a unit. In a polyloid series a diploid has two geneomes, a tetraploid four genomes, a hexaploid six genomes, etc Gametes have just half as many genomes as the zygote.

Genospecies A group, all the members of which are genotypically identical.

Genotype The genetic makeup or constitution of an individual with reference to the traits under consideration usually expressed by a symbol, e.g. "+", "D", "Dd", "sir".

Genotypic Control The control of chromosome behavicur by the genotype, in contrast to Structural control, especially at meiosis through the effects of dissimilarity of the pairing chromosomes in hybrid.

Genus A collection of similar species. Genera may be subdivided into subgenera, sections, subsections, series, and subseries. Similar genera are grouped into families.

Geoclines The graduations of variation produced within a species by adaptation within an area showing geographical variation extending over considerable distances.

Geographic Isolation Isolation of biological populations by a geographic barrier such as a mountain range a canyon, or a river.

Geographical Formation of two species from one by the interposition of a physical barrier.

Geographical Divergence The evolution from a single ancestral form of two or more different forms, subspecies or species, each in a different geographical area.

Geographical Specification The gradual formation of new species by reason of the spatial isolation of different stocks of the original species.

Geophyte A plant that produces overwintering perinating organs such as corms, bulbs or rhizomes.

Geosmin Compounds produced by some microorganisms that contributes to "earthy" smell of soil.

Germ Plasm (1) An older term for genotypes. The genetic material that is successively transferred intact from generation to generation through the germ cell. (2) The potential hereditary materials within a species, taken collectively.

Germplasm or Germ Plasm (1) The reproductive body tissues distinct from somatic [non reproductive tissues]. The genetic material, basis of heredity of an organism, passed on through previous generations.

Gerontomorphosis Evolution which is brought about by variation affecting adult organisms and which is characterised by increasing specialization and a decreasing capacity for further evolution . The production of phylogenetic effects by the modification of characters which were already present in the ancestral line of adults.

Gibberelic Acid (GA_2) A common gibberellin and one of the first to be discovered. Together with GA_1 and GA_2 a fungus that infects rice seedlings causing abnormally tall growth.

Gibberellic Acid One of the gibberellins, a group of growth hormones promoting cell division and elongation.

Gibberellin A plant hormone involved chiefly in shoot extension. Gibberellins are diterpenoids; their molecules have the gibbane skeleton. More than thirty have been isolated, the first and one of the most common being gibberellic acid, GA_3

Gigas Type An individual, biotype, or group of biotypes with "giant" characteristics; often used of describe organism with increased dimensions owing to polyploidy.

Gizzard A very muscular part of stomach of birds that grinds up food, sometimes with the aid of fragments of stone.

Globin Gene Anyone of a small family of genes in the eukaryotic genome. The genes are developmentally regulated during embryonic, foetal and adult life. Disturbances of the regulation of the globin genes result in some forms of the complex of human diseases, the thalassaemias. The genes coding for alphaglobins are clustered on chromosome 16 (in the human and most bertebrates) whereas the betablogin genes are clustered on chromosome 11.

Globular Protein A soluble pretein in which the polypeptide chain is lightly folded in three dimensions to yield a lgobular shape.

Globulin Globular protein in which the amino acid chain is folded to a compact configuraton, myoglobin, serum albumin and some of the hormone-binding proteins (e.g. thyroglobin) are globular proteins.

Gloerulonephritis Disease of kidney, characterized by inflammatory changes in glomeruli; may occur as a result of an autoimmune disease in which the streptococci induce manufacture of antibodies against the individual's own kidney tissue.

Glucocorticids Steroid hormones made by the adrenal cortex that have, among their other effects, modifications of carbohydrate metabolism. An example of a glucocorticoid is cortisone.

Glucogenic Amino Acids Amino acids whose carbon chains can be metabolically converted into glucose or glycogen.

Gluconeogenesis The formation of glucose from non-carhydrate metabolic intermediates through the reversal of glucolysis.

Glucose Oidase An enzyme that converts glucose to gluconic acid and hydrogen peroxide in the presence of oxygen.

Glucose Syrup A mixture of glucose and higher sacchardes produced by acid or enzyme hydrolysis of starch.

Glucose-6-phosphate Dehydrogenase Deficiency Commonest pathological enzyme deficiency in the human. The gene is carried on the X chromosome and females heterozygous for the enzyme can be shown to have erythrocytes wit either one or the other enzyme form. Males heterozygous for the deficiency have short-lived erythrocytes and are particularly susceptible to antimalarial drugs (e.g. Primaquine). There is, however, evidence that such males naturally possess a heightened resistance to infection by the malarial parasite.

Glucose (dextrose, grape sugar) A monosaccharide ($C_6H_6O_6$) occurring widely in nature as D-glucose. It occurs as glucose units in glycogen (hydrolysis to glucose) and sucrose (hydrolysis to glucose and fructose) as well as in starch and cellulose.

Glutamate Synthase An enzyme tat catalyze the formation of two molecules of glutamic acid from one molecule of glutamine and one molecule of A-Ketoglutarate (2-oxoglutarate).

Glutamine One of the 20 common amino acids found in proteins. It carries an acidic side-chain.

Glutathione Tripeptide coezyme comprising glycine, cysteine and glutamic acid. It serves as a hydrogen acceptorin reactions such as the formation of disulphide linkages and thiol groups in proteins.

Gluten A mixture of proteins found in wheat flour. It is composed mainly of two proteins (gliaden and glutelin), the proteins being present in almost equal quantities. Certain people are sensitive to gluten (celiac disease) and must have a gluten-free diet.

Glycerol (glycerin; 1, 2, 3-propanetiol It an alcohol with three OH groups Glycerol is biologically important as the alcohol involved in lipid formation (these particular lipids beind called glycerides).

Glycogen Storage Diseases Group of disease in which abnormally low or abnormally high amounts of glycogen are stored I the tissues. The diseases show a familial pattern of inheritance.

Glycogenolysis Breakdown of glycogen into its glucose subunits. The hormones adrenaline (epinephrine) and glucagons induce glycogenolysis in the liver via a casecade of intracellular secondary messengers.

Glycoprotein Protein in which a carbohydrate is covalently bonded to the peptide portion of the molecule.

Glycosidic Bond Chemical bond produced by the elimination of water from the anomeric hydroxyl group of an aldose or ketose and a second alcohol group to form an ether bond.

Glycosylation Attachment of an oligosaccharide to the asparagines amino group of proteins. The process is only known to take place in the proximity of the rough endoplasmic reticulum of the cell.

Golgi Apparatus A system of concentrically folded membrances found in the cytoplasm of eukaryotic cells. Plays a role in the production and release of secretory materials such as the gigestive enzymes manufactured in pancrease. First described by Camillo Golgi.

Golgi Body A complex membranous organelle or eukaryotic cells; function in secretion to the exterior and formation of the plasma membrane.

Gones The four cells or their nuclei which are normally the immediate product of meiosis.

Gonomery The state of having the maternal and paternal geneomes remaining more or less independent after the fusion of the male and femle pronuclei.

Goodpasture's Syndrome A disease of lungs and kidneys in which there is accumulation of a very uniform layer of autoantibodiers to basement membrane.

Graft Inoculation Test A viral indicator test in which a suspected viral carrier grafted to anindicator plant. If symptoms appear the viral assay is positive.

Graft Rejection The process whereby a host rejects tissues or cells derived from another organism. The mechanism of rejection is usually due to the production of specific antibodies against the new tissue which contains antigenic sites.

Gram Stain Differential stain which provides for groping of bacteria as either gram-positive or gram-negative. If the organism retains the initial stain after exposure to a decolorizer, it is gram-negative.

Gram-positive Bacteria whch resist decolourisation byacetone during the Gram's staining reaction Grampositive bacteria have asimpleer cell wall than do Gram-negative beacteria.

Granule, Metachromatic A deeply staining body in the protoplasm of some cells. Found in some bacteria and yeast.

Granuloma A nodue formed intissue as a result of the accumulation of fused macrophges, which try to destry infectious agents.

Granum A stack o membranes (resembling a pile of coins) in a chloroplast. With the light microscope these stacks are just visible as grains (grane).

GRAS A abbreviation for 'generally regarded as safe's a designation given to some food matrila, drugsetcin the USA in particular,

GRAS status may be given to compounds that are closely related to compounds that are closely related to compounds already being used for a similar purpose, or compounds that have a long history of widespread safe used although they have not been subjected to detailed testing.

Gregarious Parasitod A insect parsite which normally develops successfully at a rate of two or more individuals per arthropod host; cf Solitary Parasitoid.

Grey Crescent Band of light pigmentation that appears shortly after fertilization in the mamphibian egg. If forms opposite to the site of sperm entry and reflects a reorganization of cytoskeleton by the centriole which has been introduced by the sperm.

Grey Matter Nerve tissue tat consists mainly of nerve cell bodies and their connections, giving it a grayish colour. It occurs in the core of the spinal cord and in many parts of the brain, especially the cerebra cortex.

Group Variation Variation between different populations within a species, as distinct from "individual variation" which is the variation within a population.

Growth Curve A plot of cell number against time of a proliferating cell culture. Usually divided into log phase, before growth is initiated: log phase, the period of exponential growth; and plateau, a stable cell count achieved when the culture stops growing at a high cell density.

Growth Regulator Any organic substance other than a nutrient, which in small amounts induces a change in or otherwise influences a changes in or otherwise influences plant grow the and development.

Guard Cell One of a pair of cells that surrounds the stoma pore of a plant. The guard cells, which are kidney shaped, open and close the stoma by changes in their internal turgor pressure.

Guillain-barre Syndrome (Indry-G-B-Syndrome) Acute infective polyneuritis, associated with cytomegalovirus infection in which there is cell-mediated immunity to a component of myelin; the disease may be autoimmune in origin. Causes a temporary paralysis, particularly of the extremities.

Gymnosperm A member of a group of vascular plants of about 700 species, whose seeds are not enclosed within an ovary and in which an ovary and in which double fertilization does not occur so the seeds possess no endosperm. For example, conifers.

Gynandromorph Animals with a chimeral constitution in which certain parts of the body are female and other parts are male. The female and male parts have different constitutions as regards the sex chromosomes.

Gynergate A 'Female-Worker' type of sex-mosaic which occurs in ants. Lateral. An individual one side of which shows female characteristics and other side 'worker' character.

Gynogenesis Parthenogenetic reproduction in which the egg is stimulated to commence cleavage by the sperm but no actual fertilization has occurred.

H

H-2 Complex Mouse major histocompatibility complex, which contains immune response and transplantation antigen genes, it is located on chromosome 17.

Haberlandt, G An inspirational pioneer botanist who was the first to attempt to culture plant cells at the turn of the 20th century. He predicted the existence of hormones that would stimulate cell division (auxins) and suggested the possibility of exploiting the totipotentiality of plant cells.

Habituated Cultrue (Anergized Culture) Epigenetic change undergone by plant cells in vitro (in suspension or callus culture) that results in simplified requirements for growth. Usually the dependency upon exogenous Auxins and/or cytokinins Ceases as the cells commence sythesis of their own aendogenous supplies.

Habituation The simplest form learning, in which an animal presented with a stimulus without reward or punishment eventually ceases to respond.

Haemacytometer A microscopeslide with a grid (each square 0.05 x 0.05 mm) designed for estimating the numbers of particles insuspension, especially of blood cells. A drop of a standard dilution of the material (often stained) is placed on the grid and covered with coverslip. The suspended material can be counted rapidly an deasily. Some haemacytoeters are also designed to hold a fixed volume of floud beneath the coverslip. Haemoerythrin Respiratory pigment similar to haemoglobin, and occurring in blood corpuscles of some invertebrates. Unlike haemoglobin, haemerythrin has two iron atoms bound directly to the polypeptide chains of the molecule.

Haemocyanin A blue copper-containing blood pigment founding many mollusks and arthropods. Haemocyanin is the second most abundanet blood pigment after haemoglobin and functions similarly in acting as an oxygen carrier in the blood.

Haemoglobin Protein carrier of oxygen found in red blood cells. Composed of two pairs of identical polypeptide chains and an iron-containing heme group.

Haemolytic Anaemia Disorder in which the amount of haemoglobin per unit volume of bloods is decreased by actual red ell destruction.

Hairpin Loops Regions of double helix formed by the pairing of two continuous complementary streches of bases on the same single DNA or RNA strand.

Half-life, Radioactive Time required for a unit mass of a radioisotope to lose one-half of its total original radioactivity due to decay into another non-radioactive isotope. Halflives of radio nuclides vary greatly, being a few days for 32P, over 12 years for ^{3}H and 5700 years for ^{14}C.

Half-tetrad Analysis Analysis of chromosome tetrads following recombination where only two out of the original four chromosomes of the tetrad can be analysed. The attached X chromosomes of drospphila provide the best known example.

Hanging-drop Technique A technique for observing microorganisms suspended in a drop of fluid.

Haplo A plant in which there is no alternation of generations so that there is only one type of individual in the complete life-history.

Haplodization Process whereby what was originally a diploid nucleus becomes haploid, presumably by a progressive loss of one member of each chromosome pair.

Haploid (monoploid) An organism with only one geneme in the cell nclei. In so-called polyhaploids (Individuals with halved chromosome numbers derived from polyploid species), there are, however, more than one genome. The term haploid is also used to indicate that gametes (in diploid species) only contain one chromosome of each kind. Thus, the gametes are haploid; and the somatic cells are diploid.

Haplophase Stage of development in different organism having the same number of chromosomes in the cell nuclei as in the gametes. This number is half that occurring in the diplophase. As a rule, each chromosome type in the haplophaseis represented by only one chromosome.

Hapten A substance that by itself does not elicit an immune sponse unless conjugated with an immunogenic carrier molecule; may react specifically with immune products in the sense of a carrier molecule.

Hartnup disease Amino acid transport defect that leads to excessive loss of monoamino monocarboxylic acids (cyrtine, lysine, ornithine, arginine) in the urine, and poor absorption in the gut.

Harvey Sarcoma Virus Mouse retrovirus carrying the V-ras oncogene which is homologous to the cellular protoonocogene C-ras, which is located on chromosome 11.

HAT Selective Medium Cell culture medium containing hypoxanthine, aminoperin and thymidine used to select hybridomas for the production of monoclonal antibodies. Myeloma cells used in the fusion process cannot grow in HAT mediumunless fused with a spienocyte cell to provide salvage pathway enzyme for nucleotide biosyntesis. Such selective media select cells able to produce HGPRT and thymidine kinases, and are also used to select other types of fused cell constructs beside hybridomas.

Haustorium A specialized outgrow of certain parasiticplants, .g. fungi that pentrates into and withdraws food material from the cells of the host plant.

HCRM, Host Controlled Restriction and Modification The mechanism by which certain bacteria attempt to protect themselves from phase infection. Incoming phage (foreign) DNA is destroyed by site-specific restriction

Heat Exchanger A mechanical device used to bring two fluids of different temperature into close contact so that heat may be efficiently transferred from the fluid at the higher temperature to that at the lower temperature.

Heat Shock Proteins Group of proteins that are synthesized following heat shocking of a cell or tissue and probably help to protect the cell against some of the deleterious effect of heat shock.

Heat hock Puff Gene activation in response to heat shock in Drosophila larvae leads to the appearance of these puffs in the polythene chromosomes. Such puffing represents the transcriptional activation of the heat shock genes in the larvae.

HEAVY Isotopes Forms of atoms containing greater than the common number of neutrons and thus more dense than the commonly observed isotope (e.g. ^{15}N, ^{13}C).

HEINZ Body (inclusion body) Agregate, precipitated globin molecules, found intracellularlyand visible by light microsocopy, they are the result of the dissociation of methaemoglobin.

HeLa Cells An established line of human cervical carcinoma (cancer) cells. Has been used for many years in study of biochemistry and growth of cultured human cells.

Helper Cells A functional subclass of T cells which can help to generate cytotoxic t cells and cooperate with B cells in the production of an antibody response. Helper cells usually recognize antigen in association with class 2 MHC molecules.

Helper Factor a group of factors apparently produced by helper T-lumphocytes which act specifically or nonspecifically to transfer T-cell help to other classes of lymphocytes. The existance of specific T-cell helper factor is uncertain.

Helper Virus A defective phage which can help another defective phage (q.v.) to make a successful infection, cf. Complementation.

Heme Protein A protein containing a heme as a prosthetic group.

Heme The iron-porphyrin prosthetic group of heme proteins.

Hemicellulose Group of alkali-soluble polysaccarides that are important constituents of plant cell walls and also act as food reserve in some plants. They include mannans, glucans, xylans and galactans.

Hemidesmosome Cellular structure morphologically resembling half a desmosome occurring on the basal aspect of epithelial cells and assumbed to serve a comparable adhesive function, in this case, anchoring the cells to the basal lamina. Recent evidence indicates certain compositional differences from true desmosome but these may turn out to be minor.

Hemoglobin The oxygen-carrying pigment of erythrocytes, having the reversible property of taking up oxygen (oxyhemoglobin, HbO_2) or of releasing it (reduced hemoglobin, Hb), depending primarily on the oxygen tension of the medium surrounding it.

Hemophilia A tendency to profuse bleeding even from slight wounds. Inheritance I governed by a recessive gene in the nonhomologus portion of the X chromosome in man. Hemophilia is a sex-inked abnormality.

Hemosiderin An iron-containing glycoprotein pigment found mainly in the liver, spleen, and bone, marrow, representing colloidal iron in the form of granules larger than ferritin molecules.

Heparin A substance that prevent blood clotting by neutralizing prothrombin and stopping the action of thrombin. It is present in tissues of mammals, and is secreted by some blood-sucking animals.

Hepatic Portal System Avenous pathway comprising the hepatc portal vein, which carries blood rich in absorbed food materials such as lucose and amino acids, from the intestine to the liver. There the materials may be stored, converted, or released to the general circulation via the hepatic vein.

Hepatocyte Main cell type of the liver, derived from primitive gut epithelia. Hepatocytes are arranged in folded sheets facing the blood-filled sinusoids.

Herbicides Resistance Resistance or tolerance to herbicides. This is a field of intensive interest among those involved in plants tissue culture work. Screening or selection at the celular level for putative resistance is tested by plant regeneration and followed by additional screening of these plants and their progeny.

Hereditary Angioedema A condition caused by congenital deficiency of C1 INH and a pathological activation of the classical complement pathway.

Hereditary Disease A pathological condition whose causes is a gene mutation and that can therefore be transferred from one generation to the next.

Hereditary Univalents Chromosome which has evolved a system of more or less even distribution in segregation without going through the process of pathytene pairing cf Spontaneous Univalent.

Heredity The occurrence of identical characteristics in oganism belonging to different generations. Also the fundamental biological causes of this identity.

Heritability That part of the variance of certain quantitative character that is due to genetic causes, of this identity.

Heritability That part of the variance of certain quantitative character that is due t genetic causes, heritability is calculated by means of statistical methods and presupposes that the character in question is conditioned by polymeric genes.

Hermaphrodistism The bearing of effective male and female gametes by the same flower in Phanerogams, by the same flower in Phanerogams, by the same flower in phanerogams, by the same prothallus in crystogams and by the same individual in animal can be simultaneous or consecutive in animals. Sometimes in correctively used for intersexuality.

Herpes An enveloped animal virus of –200nm diameter, which posses a linear duplex DNA molecule within an icosahedral capsid. Herpes-type viruses have been implicated as the causative agents of cold scores, shingles and mononucleosis.

Heterauxesis The relation of the growth-rate of part of developing organism (whether morphological or chemical) to the growth-rate of the whole or of another part, a comparison between organism of the same group but of different ages and hence sixes.

Heterochromatism Appearance of different colours in the flowers of the same inflorescence due to seasonal differences.

Heterochromosomes Cromosomes distinguished by special peculiarities of behaviour, form or size in contradistinction to autosomes or euchromosomes.

Heteroecy Dicey Correns (1928). Subdivided by Hartmann (1981) into Diplo-heteroecy. Diploid dioecy and Haplo-heteroecy. Haploid dioecy.

Heterogametic A term denoting that one of the sexes forms two kind of gametes with regard to sex determination, e.g. With X or Y chromosomes. The sex that forms only one type of gametes with regard to sex determination is said to be homogametic.

Heterogenetic Association The pairing in an allotetraploid, of chromosomes derived from different ancerstral stocks.

Heterogenetic Pairing Pairing and subsequent segration between chromosomes derived from the two parental species from the two parental species of an allopolyploid.

Heterogenote Patilly diploid (merozygous) bacteria which are heterozygous with regard to loci represented on both the exogenote and the endogenote.

Heterogenous Nuclear RNA (hnRNA) Complex mixture of RNA molecules of varying sizes up to at least 30,000 nucleotides long found in the nuclei of eucaryotic cells, hnRNA molecules are synthesized at a rapid rate and have short halflives, generally between 0 to 30 minutes in vivo. They include the largest types of RNA known in nature. Transcription of hRNA is by RNA polymerase 2, and the nascent transcripts are packaged with proteins and small nuclear RNA (snRNA) into ribonucleoprotein (RNP) particles as they are produced.

Heterogenous With more than one type of cell or individuals is callus, which is usually non-uniform in cell composition.

Heterogeonic Growth The growth of two part, or organs, at different rates, which however, keep a constant ratio to each other.

Heterograftor Xerograft An interspecific graft. The likelihood of success of this graft is proportional to the degree of relatedness between the donor and recipient.

Heterokaryon A multinucleate cell containing nuclei of more than one genotype, such as are of widespread occurrence in fungi.

Heterokaryosis The presence of genetically different nuclei within individual cells of mycelium.

Heteromorphism The existence of more than one from, used especially with reference to life cycles in which the alternating generations are markedly different morphologically, as in ferns and jellyfish, compare Isomorphism.

Heterophil Antibody An antibody that reacts with totally unrelated species of microorganisms or cells from unrelated species of animals. Agglutination of Proteus spp cells by serum from typhus fever patients is an example.

Heterophile Antigen A substance preset in various tissue that causes the production of immunoglobulins which react with the tissues of several mammals fish and plants.

Heterophylly Denoting plant species that have more than one form of lfoliage leaf on the same individual. Where young leaves differ from adult leaves as in Sagittaria, the phenomenon is termed environmental heterophylly. A few species exhibit heterophylly at random.

Heteroplasmon or heteroplast A cell with a mixture of two types of cytoplasm [cytoplasmic hybrid]; cell with foreign organells.

Hteroplastic Graft A graft or transplantation of tissue, between different species or between different genera as opposed to a homoeoplastic graft.

Heteroplastic Transplanstation Transplantation of tissue between individual of the same genus but differing species.

Heteroycnosis Excessive charging of heterochromation with nucleic acid in meiotic and premeiotic divisions Negative Hetropycnosis, deficient charging of heterchromatin with nucleic acid starvation of Darlington and La cour (1940)

Heterosis The increased vigour often exhibited by hybrid individuals.

Heterospory In higher plants the production of two kinds of mmelocytes (megasporocytes and microsporocytes), giving rise to two different kinds of meiospors. Compare with Homospory.

Heterostyly The phenomenon in higher plants in which the style in the flower is of a different length in different individuals, generally the stamens also are attached at different leaves in the flower, or are of different lengths. This mechanism favours ross-fertilization between individuals differing in this kind of flower structure.

Heterothalic Yeasts Yeasts with complex sexual compatibility systems in which specific mating types are required.

Heterotroph A microorganism which is unable to use carbon dioxide sits sole source of carbon and which requires one or more organic compounds.

Heteroxenous Describing a species which requires the use of more than one host species to successfully complete its annual life cycle.

Hetrozygosity Index Measure of the number of loci for which an individual is heteroxygous.

Heterozygote (adj) Heterozygous An individual whose chromosomes bear unlike genes of a given allelic pair or series. Heterozygotes produce more than one kind of gamete with respect to a particular locus.

Heterozygotes, Proportion of In a population mating at random the proportion of individuals heterozygous for any particular gene can be found from the equation $p^2:2pq:q^2$ where p^2 and q^2 are the proportions of the two homozygous classes respectively, 2pq represents the proportion of heterozygotes, and $p + q = 1$.

Hetrokinesis The meiotic division during which gametes of different sex-potentially becomes separated by differential distribution of the sex chromosomes.

Hexad Unequal bivalent (not trivalent) formed by pairing of one chromosomes with its homolous which has fused with a third non-homologus chromosome, in a fusion hetrozygote.

HFR (High Frequency of Recombination) Strains of E. Coli that show unusually high frequencies of recombination. In these cells the F factor is integrated into the bacterial chromosomes, where it is brought to play some part in the transfer of the chromosome from HFR to F-cells.

HGPRT (Hypoxanthine Phosphoribosyl Transferees) Enzyme that catalyses transfer of the phoshoribosyl moiety to hypoxanthine and guanine.

High Added Value Descriptive of any process in which the economic value of a product is many times the cost of the raw material required for its production.

High Fructose Syrup A syrup composed of the hexose sugars glucose and fructose in roughly equal amounts. High fructose syrups are produced from starch using enzymes.

High Rate Biological Filter An advanced technique for the arobic treatment of wastewater and sewage. In such filters the traditional gravel is replaced by specially designed plastic media, often in the form of rings.

High-efficiency Filter Having a nominal efficiency of 99% for removal of 0.3 to 0.5 micron particles from air.

High-energy Compound A compound that undergoes a large decrease in free energy upon hydrolysis under standard conditions.

High-mobility Syrup Production High fructose syrups are produced by the action of glucose isomerase on a syrup contain in a high glucose level, derived by hydrolysis of starch.

Highly Repetitive DNA Sequence of DNA that occurs repeatedly in a given genomic sample of DNA. Such DNA is chiefly centrometric and plaindromic DNA. DNA reassociation and cost plot analysis real the respective amounts of repetitions DNA in a sample.

Hill Reaction The oxygen evolution and photoreduction of an artificial electon acceptor by a chloroplast preparation in the absence of carbondioxide.

Hinge The portion of an immunoglobulin heavy chain between the Fc and Fab regions. This allows flexibility within the molecule and allows the two combining sites to operate independently. The hinge region is usually encoded by a separate exon.

Histocompatibility The ability to accept grafts between individuals

Holoblastic Describing the type of cleavage seen in certain eggs in which the first cleavage (animal-to vegetal in orientation) divide the egg cytoplasm and yolk completely, as in sea urchins and frogs. Sometimes as in most yolk is cleaved this is not hololastic cleavage.

Holoenzyme A catalytically active complex made up of an apoenxyme and a coenzyme. The former is responsible for the specificity of the holoennzyme whilst the latter determines the nature of the reaction.

Holt-Oram Syndrome (hand-heart syndrome) Human congentital disorder caused by an autosomal dominant gene, in which arable arm malformation occurs together with atriai septal defects in the heart.

Homeostais The maintenance of a constant internal environment by an organism. It enables cells to function more efficiently. Any deviation from this balance results in reflex activity of the nervous and hormone systems, which tend to negate the effect. The degree to which homestasis is achieved by a particular group, independent of the environment, is a measure of evolution advancement.

Homoeosis (heteromorphosis) The alteration of one organ of a segmental or homologuous series from its own characteristic forms to that of some other member of the series.

Homogenetic Association or Pairing Pairing between truly homologus chromosomes, particularly used in allopolyploids of pairing between chromosomes derived from one of the parental species.

Homogenotisaton A genetic technique where one copy of a gene, or other DNA sequence within a genome, is replaced with an altered copy of that sequence. The DNA is first cloned and then altered in some way, e.g. a transposone is inserted into a gene. The mutated gene copy can be used to replace the original gene by recombination in vivo. The incorporation of the mutated gene is usually selected, e.g. by virtue of its containing a transposon-encoded antibiotic resistance.

Homogentisic Acid A reducing substance excreted in the urine of persons with a genetic metabolic deficiency known as alkaptonuria. Upon oxidation this acid turns black. Also called Alkapton.

Homologus Population A culture in which the cells are all identical (morphologically and genetically).

Homologous Of chromosomes or chromosome segments that are identical with respect to their constituents genetic loci and their visible structure; in evolution, of genes and structures that are similar in different organisms owing to their having inherted them from a common ancestor.

Homologus Chromosomes Chromosomes occurring in pairs, one derived from each of two parents, normally (except for chromosomes associated with sex) morphologically alike and bearing the same gene loci, Each member of such a pair is the omolog of the other.

Homologus Protein A protein having identical functions and similar properties in different species; e.g. the hemoglobins.

Homology Residual A degree of affinity still existing between once truly homologus chromosomes, which will now only rarely permit crossing over between them.

Homomorphic of chromosomes pairs; morphologically, in-distuishable. Opp Heteromorphic.

Homonuclus, Homunuculus The miniature foetus erroneously thought to be present in a human spermatozoan.

Homoosis The production in one particular region of a individual, of an appendage normally occurring in a different region of the individual's body.

Homopolymer Polymers which contain the same monomer units. A nucleic acid chain containing only one sort of nucleotide, e.g. the RNAs, polyuridylic acid.

Homospory In plants, the production of but one kind of meiotic, which gives rise into meiospores morphologically indistinguishable from each other.

Homozygosity The condition of beinghomozygous the proportion of homozygous individual at a locus or of homozygous loci n an individual.

Homozygote (ad) Homozygous) An individual forming only one kind of sex cell due to the fact that corresponding loci in homologus chromosomes contain identical alleles.

Homozygous Frequency The frequency with which one allele of a heterozygous gene-pair becomes homozygous in the succeeding generations the gene-pair concerned being situated in attachment X-chromosome.

Homozygous Sex The homogametic sex; which produces gametes of only one kind with regard to their properties of sex determination.

Hops The female flowers of the vine Humulus lupulus which are used in the production of the bitter flavour in beers. The essential oil from hope, which gives them their aromatic properties, mainly terpenes, and comprises about 1% of the total dry weight. The bitter components are resins.

Horizontal Transmission Transfer of genetically material byinfection-like processes, including transduction. Compare Vertical transmission.

Hormonal differences in Horn Production in Sheep In some sheep hybrids between horned and hornless breeds, the heterozygous male has horn, while the heterozygous females are hornless, presumably an influence of the male and female sex hormones.

Horseradish Periodxidase Enzymes used to identify endocytotic vesicles since the enzyme reacts with the products of the reaction are readily stained by osmium tetroxide for electron microscopy.

Host Induced Modifications Generally non-hereditary attractions in the nature of a virus induced by its host. For example, conversion of non-virulent phages into virulent ones after they pass through a host with glucosylating enzyme. The viral nucleic gets glucosylated (q.v) possessing bacteria.

Host Range The number of different species or host on which a plasmid phase or other virus will replicate.

Host Resistance Here in the sense of painter meaning he relative genetic ability of a certain variety of plant (or animal) to produce a large crop of good quality than do ordinary varieties at the same level of pest infestation: practically speaking, based on three elements: non-preference, antibiosis, and/or tolerance.

Host Restriction Bacteria of different species have a well-defined defence system against invading foreign DNA. The invaded cells have specific endonucleases which cleave this foreign DNA at specific target sequences, thereby making it susceptible to other non-specific nucleases.

Host-cell Reactivation The repair of damaged DNA of viruses by host-cell repair mechanisms.

Housekeeping Gene Gene that is expressed in most or all cells of a complex multicellular organism; the products of such genes are essential enzymes or structural proteins in the basic form and general maintenance of the cell Non-housekeeping gene are known as cell-specific genes or luxury Genes.

Human T Cell Leukaemia Virus (HTLV) Anyone of a group of retroviruses of which one, now termed human immunodeficiency virus (HIV) and previously known as HTLV3 or LAV (lymphadenopathyvirus), is the cause of acquired immune deficiency syndrome (AIDS).

Humoral Immune System The part of eh immune system mediated by B cells; it is mediated by circulating antibodies and is active against extra cellular bacterial and viral infections.

Humus A dark mass of decayed animal or plant matter remaining after microbial decomposition that gives soil a loose texture and brown or black colour.

Hunter Syndrome One of a group of genetically inherited disorders of the human involving malfunction of mucopolysacchyaride storage in connective tissue; a type of gargolism. The gene of Hunter syndrome is X-linked. Compare Hurler Syndrome.

Huntingdon's Syndrome One of a group of genetically inherited disorders of the human involving malfunctions of mucopolysaccharide storage in connective tissue; a type of gargoylism. The gene of hunter syndrome is X-linked. Compare Hurler's syndrome.

Hurler's Disease Autosomal recessive storage disease in which A-iduronidase is absent, leading to accumulation of herparan and dermatan sulphates. Extensively deposits of mucopolysaccharides are found in gargoyle cells, and in neurons.

Hyaluric Acid A type of organic acid that has the properties of a lubricant. It is found, for example, in the synovial fluid of joints.

Hyalyronidase "Spreading factor"; enzyme capable of degrading hyalyronic acid, the intracellular material of connective tissues; may help the spread of the invading organism in he tissues.

Hybrid Cell The term used to describe the mononucleate cell which results from the fusion of two different cells, leading to the formation of a synkaryon.

Hybrid Corn Northocline; an extension of the term cline to cover the gradation of forms which occurs in nature following interspecific hybridization.

Hybrid Protein A polypeptide forced by the expression of a hybrid gene.

Hybrid Selection The process of choosing plants possessing desired characteristics from among a hybrid population; identification and isolation of those desirable for further studies, propagation, etc.

Hybrid Vigour The unusual robustness, rapid growth and thriftiness of organism produced by crossing two less vigorous parents heterosis.

Hybrid, Hetrozygote or Heterozygous Usually used where chromosomes or multiple gene differences are concerned.

Hybrid-released Translation Method in which cloned DNA is bound to a cellulose nitrate filter and hybridized with a mixed population of mRNA or total cellular RNA. Hybridised message is then eluted and translatied in a cell-free translation system. Compare Hybrid-arrested translation.

Hybridisation (molecular) The matching of complementary DNA-

DNA or RNA-DNA strands to determine the number and/or location of specific nucleotide sequences.

Hybridisation, Introgressive Hybridisation in which infiltration of the germplasm of one species into another occurs; typically such infiltration is brought about by chance back crossing.

Hybridogenous Pseudo-parthenogenesis Seed production following stimulation of pollen the male nucleus of which becomes eliminated before fertilization so that a maternal embryo is formed which as a result of doubling is diploid.

Hydatidiform Mole Placental tumour resulting from excess growth of chronic villi and with a significant incidence of malignancy.

Hydrogen Peroxide Used as a disinfecting agent for plant material and as an oxidisting or bleachling agent.

Hydrolase An enzyme that catalyse a hydrolysis reaction. Digestive enzymes are an example. Hydrolases play an important part in rendering insoluble food material into a soluble form, which can then be transported in solution.

Hydrophobic Molecules and amino acid side chains, which are mainly hydrocarbons (compounds of C and H with no charged groups or polar groups), have a lower energy when they are clustered together than when they are distributed through anaqueous solution.

Hydroxyprogesterone A hormone secreted by the corpus luteum in the ovary and controlling events in pregnancy.

HyHypotroph An obligate intracellular parasie, including viruses and other forms of microorganisms requiring living cells for their nutrition.

Hyperimmune Serum Serum prepare from animals that have recent received repeated injections or applications of a choosen antigen; thus the serum should contain a very high concentration of polyclonal antibodies against that antigen.

Hypermetamorphosis A type of parasitic insect life cycle involving transformation through at lest two distinctly different larval types, especially an active host-seeking first instar and several more passive parasitic instar.

Hypersensitivity Increased sensitivity, or allergy, to foreign material. It is a state of overresponsiveness by he immune system to an antigen.

Hypertonic Designating a solution with an osmotic pressure greater than that of a specified other solution, the latter being hypnotic. When separated by a semipermeable membrane water moves by osmosis into the hypertonic solution.

Hypochromic Effect A decreases in the absorbance of ultraviolet light at 260 nm associated with the joining of free nucleotides into single-stranded nucleic acid or the formation of double stranded DNA.

Hypogammaglobulinaemia (USA hypogammaglobulinemia) Syndromes in humans and other vertebrates in which the immunolglobin level is depressed below the normal range. Congenital, chronic and transient types are known.

Hypomorphic (mutation) A gene mutation producing an effect similar to but to as great as, that produced by agene loss.

Hypothalamus The part of the brain lying belong the thalamus; it coordinates water balance, reproduction, temperature regulation, and metabolism.

Hypotonic Designating a solution with an osmotic pressure less than that of a specified other solution, the latter being hypertonic. When separated by a semipermeable membrane water I lost by osmosis from the hypotonic to the hypertonic solution.

I

I Band One of the bands appearing in light microscope sections of striated muscle fibre. It represents filament of action, troponin and tropomyosin which overlap with the filament of the A BAND.

I-cell Disease A human disease in which the lysosomes lack hydrolases but high concentrations of these enzymes are found in the extra cellular fluids. The gene defect responsible probably prevents the addition of the lysosome recognition marker (mannose-6 phosphate) to these enzymes so that they are not directed into the lysosomes but are released.

Iatrogenic A disorder or disease caused by the diagnosis or treatment by a physician.

ICAM (Intracellular Cell Adhesion Molecule) ICAM-1(90kD) is the ligand for the leucocyte integrin LFFA-1. It is expressed on fibrolasts andendothelial cells, and its levels can be up-regulated by lymphokine treatment. One of the immunogloblulin superfamily.

Ideal Free Distribution Distribution of organisms resulting from a process in which each newly arriving individual selects the habitat that is best for its success. (Contrast with dominance distribution.)

Identical in Structure Of two genes that are identical in nucleotide sequences regardless of whether or not they are both derived form a common ancestor.

Idioblast Any specialisation plant cell that is dispersed among cells of a different kind. Idioblat may contain a variety of material, e.g. tannins, oils, crystals, and waste products. Isolated sclereieds can be called idioblast.

Idiogram Diagrammatic representation of the chromosome set arranged according to size, shape and the position of the centromere, Generally speaking, the largest chromosome is placed at the extreme left and the smallest one at the right end with the intermediate forms integrading between the two extremes.

Idiotroph A microorgnism that has amutation in its biosynthetic pathways associated wit the production of secondary products, for instance, in the production of antibiotics such mutants are not capable of making one part of the antibiotic. Hence for antibiotic production to take place the medium must be supplemented with the missingmoiety. When the mutant is fed analogues of the missing part new antibiotics are produced.

Idiotypic Markers Those antigenic determinants in the antigen binding of an immunoglobulin which stimulate idiotypic antibody formation.

Idiozome The zone of non-granular protoplasm around the centrosome; the attraction sphere.

IgM (macroglobulin) Immunoglobulin of the primary immue reaction mammals. (Mr900000), its sedimentation coefficient is 19S, and its carbohydrate content is 12 percent. IgM consists of a pentamer of five basic immunoglobulin units.

IgX An immunoglobulin class found in a Amhibia.

IgY Low-molecular-weight (sedimentation coefficient 7S) immunoglobulin of birds, reptiles and frogs and toads.

Illegitimate Recombination Recombination between DNA species which show very little or on homology. Illegitimate recombination is promoted by transposons and IS elements.

Illuminate To supply or brighten with light. Illumination is an absolute requirement for most kinds of plant tissue cultures. Fluorescent lights are most commonly employed, the intensity of which is dependent on the light source and the requirement of the culture. The process is illumination which is also the total visible radiation on a surface.

Imaginal discs Groups of cells found as single-cell layers of infolded tissue beneath the epidermis of holmetabolous insects (insects having a series of distinct stages in their metamorphoasis).

Imaginal Tissue Undifferentiated tissue in the larva or embryo from which the external organs of the final instar (the imago or fully adult insect) develop.

Imbibition The phenomenon in which a substance absorbs a liquid and swells, but does not necessary dissolve in the liquid.

Immediate Hypersensitivity Reaction involving antibodies and occurring within a few minutes of exposure of the individual to the antigen or allergen. Individuals displaying such hypersensitivity have normally been previously sensitized to the antigen.

Immortalised Cell Cell which by dint of transformation, by infection with virus, becomes capable of indefinite growth and division. Established cell lines in tissue culture are composed of immortalized cells.

Immotile Cillia Syndrome Cogneitaldefect in dynein (either absent or inactive0 which leads to male sterility and poor bronchial function. Interstingly, non-cilliated cells show altered locomotion and 50% of patients have Kartagener' syndrome.

Immune Competent Cell (immunocompotent cell) Cell that is capable of responding to an antigen, by division, transformation and production of antibodies or lymphokines or acuisition of other immunological effect or function. Small lymphocytes are normally, immunocompetent in vertebrates. When stimulated by antigen, B lymphocytes divide and transform into antibody-secreting plasma cells; T lymphocytes divide and transform into effector cells responsible for cell mediated immunity and t-B cell cooperation.

Immune Complex, Disease Disease characterised by the presence of immune complexes in body fluids. Hypersensitivity of the earths reaction type and serum sickness are examples.

Immune Modulator It is the compou.. ᵈ which interferes with the production of an antibody in response to an antigen. Immune modulators are used to reduce the possibility of a host rejecting a hetrograft.

Immune Reaction Measurable r observable reaction or response arising from antibody-antigen interaction, or from contact of immune competent cells with antigen. Response and reactions (e.g., cell division, production of an antibody in response to an antigen. Immune modulators are used to reduce the possibility of a host rejecting a hetrograft.

Immune Reaction Measurable or observable reaction or response arising from antibody-antigen interaction, or from contact of immune competent cells with antigen. Response and reactions (e.g., cell division, production of antibody, release of lymphokines and histaine, induction of immunological tolerance and allograft rejection) can occur in viro and/or in vivo.

Immune Surveillance Postulated mechanism hereby abnormal cells containing new antigen that have arisen through somatic multation are removed from the body as a result of continuous monitoring of body cells by the immunological system. It has been suggested that potential cancer cells are removed in this way.

Immunity The ability of plants and animals to withstand harmful infective agents and toxins.

Immunization Method of imparting or injecting an immunogen (q.v.) into a susceptible host organism.

Immunoadsorbent Any insoluble, e.g. celluose, with either an antigen or an antibody bound to it and which will bind its corresponding antibody or antigen thus removing it from a solution.

Immunoaffinity Similarity in the immune properties of two or more substance; in mechanism of attraction between an antibody and an antigen.

Immunoassay An assay system which detects proteins by using an antibody specific to that protein. A positive result is seen as a precipitate of an antibody – protein complex. The antibody can be linked to a radioactive atom or to an enzyme which catalyses an easily monitored reaction.

Immunocytochemical Method A method that uses a singal compound linked to specific antibody to locate a portion in a cell.

Immunocytochemistry Techniques for staining cells or tissues using antibodies against the appropriate antigen.

Immunoeletrophoresis An analytical technique used for the identification of antigen separated from a mixture of proteins (e.g. blood serum) by electrophoresis.

Immunofluorescence Flurescence resulting from a reaction between a substance and specific immunoglobulins that are bound to a fluorescent dye.

Immunogenetics The study of gene controlled synthesis of antibodies and their reaction with antigens.

Immunoglobulin Gene Gene that encodes that light chains and heavy chain of the immunoglobulin molecules. The great variability of antibody structure is achieved by a series of remarkable genetic mechanisms, which involve rearrangement of parts of the DNA sequence during the life of the B lymphocyte. The variable and constant regions are coded separately on the genome, but in a B lymphocyte they are made adjacent, with an additional J or joining sequence between. Both genetic rearrangement and somatic mutation are involved in increasing the sequence diversity of the V region genes.

Immunological Memory The systems responsible for the situation where reactions to a second or subsequent exposure to an antigen are more extensive than those seen on first exposure. The memory is best explained by colonal expansion and persistence of such clones following the first exposure to antigen.

Immunological Network The concept due to Jerne that the entire specific immune system within an animal is made up of a series of interacting molecules and cell surface receptors, based on the idea that every antibody combining site carries its own marker antigens or idiotypes and that these in turn may be recognized by another set of antibody combining-sites and so on.

Immunological Tolerance Absence of immune response to antigens, resulting from recognition of "self" or induced by very large antigen dose.

Immunosuppression The reduction in the capability of the immune system to respond to an antigen; obtained through immunosuppressive drugs irradiation, and antilymphocytiscerum (ALS).

Immunotoxin An artificial product obtained by fusing an antibody with a toxin of plant or bacterial origin.

Imperfect Fungi Fungi that do not have a sexual cycle.

Implantation Introduction of tissue grafts to a recipient animal. (2) Attachment of the fertilized ovum or mammalian embryo to the endometrium of the uterus of the mother. Synonymous with Nidation.

Impregnation The deposition of spermatozoa within the body of the female whether brought about by coitus or artificially; insemination.

Imprinting A form of learning in the first few hours of a young animal life in which it fixes its attention on the first object with which it has contact (visual, auditory, tactile, or olfactory) and afterwards follows that object.

Imunoprecipitation The precipitation of a multivalent antigen by a bivalent antibody, resulting in the formation of a large complex. The antibody and antigen must be soluble. Precipitation usually occurs optimally when there is near equivalence between antibody and antigen concentration.

In Situ Hybridization Molecular hybridization carried out using unfractonated cells or cell organelles. Radioactive orenzymically labeled probes are added to a cell squash on a microscope slide and allowed to hybridize to sequences in intact chromosomes. In this way the precise locations of genes and other DNA sequences in the genome can be determined. In this technique a labeled DNA probe is hybridized to metaphase chromosome. The position of hybridization is visualized at the light microscope level.

In Vitr Fertilization Fertilization of an egg outside the body of the female and in the test tube', using sperm already obtain from a male of the same species. Such fertilization is widely used to overcome human infertility due to blockage of the fallopian tubes, the resulting embryo being subsequently introduced into the uterus of its original donor.

In Vitro Transformation A heritable change, occurring in cell in culture, either intrinsically or from treatment with chemical carcinogens, oncogenic viruses, irradiation, etc., and leading to the acquisition of altered morphological, antigenic, neoplastic, proliferative, or other properties.

In vivo (Latin:in Life Pertaining to experiments done in a system such that the organism remain intact, either at the level of the cell (for bacteria) or at the level of the whole organism (for animals).

In vivo Culturing of Imaginal Discs Procedure in which the imaginal discs of Drosophila larvae may be cultured by injection into the abdomen of a host larva or adult. This has permitted the distinction to be drawn between the stages of determination and differentiation in such tissues.

In Vivo Marker Mutation that occurs in the organism, but also allows detection in tissue culture; for example, genes causing galactosaemia, being detectale both in the phenotype of an individual and in vitro in tissue cells derived from such individuals.

Inborn error of Metabolism Metabolic deficiencies inherited in Mendelian manner and produced by recessive genes; a term coined by Sir A.E. Garrod to apply to biochemical mutants in man.

Inbred Line A line showing a considerable degree of homozygosity as a result of continued inbreeding and selection.

In bread Strain Any strain of animal or plant obtained by a breeding strategy which tends to lead to homozygosity. Such breeding strategies include brother-sister mating and back-crossing of offspring with parents.

Inbreeding Coefficient The probability that the two genes 9allles) at a locus are identical by descent.

Inbreeding Minimum The stage reached after a long period of inbreeding. Because of the homozygosity attained, vigour is not further decreased but remains contain if inbreeding I continued.

Incest (in man) Breeding between individuals related in degrees prohibited by custom, tabu or canonical injunction.

Incineration The process of burning to ashes; e.g., sterilizing the inoculating loop in the Bunsen burner.

Incipient Plasmolysis The condition of a cell that results when it is surrounded by a solution having the same osmotic pressure as the cell contents.

Incipient Species Population that are too distinct to be considered as subspecies of the same species, but not sufficiently differentiated to be regarded as different species; also called Semispecies.

Inclusion Bodies General name given to the characteristic particulate structures present in certain virus infected host cells. For instance, the Negri bodies present in nerve cells infected by tabbies virus. These bodies act as diagnostic features.

Inclusive Fitness The sum of an individual's own fitness (the effect of producing its own offspring; the individual selection component) plus its influences on fitness in relatives other than direct descendants (the kind selection component).

Incubator An apparatus providing controlled environmental conditions, (light, photoperiod, temperature, humidity, etc.} suitable for (incubating) plants or plant cultures. This term is sometimes used as synonym for cabinet (growth cabinet) but usually implies a greater degree of environmental control.

Independent Assortment Genes located in different chromosomes are generally inherited independently of each other. This process of inheritance is called independent assortment, a phenomenon discovered by Mendel.

Independent Comparisons Comparisons between observations whose value are uninfluenced by changes in each other. Such comparisons are termed Orthogonal and functions from which they are calculated are Orthogonal Functions (such as are used in manipulating genetical data for the detection of linkage). Indexing Any of several procedures for demonstrating the presence of specific viruses in suspect plants. It depends upon serological assay or the transfer of some part of this plant (such s buds, scions, sap, etc.) to an indicator plant.

Indexing Any of several procedures for demonstrating the presence of specific viruses in suspect plants. It depends upon serological assay or the transfer of some part of this plant (such as buds, scions, sap, etc.) to an indictor plant.

Individual Selection Component That component of inclusive fitness that results from an organism producing its own offspring.

Inducible Enzyme An enzyme that is present in much larger amounts when a particular compound (the inducer) has been added to the system.(contrast with constitutive enzyme).

Infectious Microorganisms The term is restricted to microorganisms capable of producing a disease process in man or domestic animals.

Innate Releasing Mechanism (IRM) A system within the central nervous system that is especially responsive to a particular sign stimulus.

Insertion Element Generic term for DNA insertion sequences found in bacteria capable of genome insertion. Postulated to be responsible for site-specific phage and plasmid integration.

Intention Movement The preparatory motions that animals go through prior to a complete behaviour response; for example, the crouch before flying, the snarl before bitting etc.

Interference Competition Competition resulting from direct behavioural interactions between organisms.

Intrinsic Protein A membrane protein that is embedded in the phospholipid bilayer of the embrane.

Intron A protein of a DNA molecule that, because of splicing out, is not

Ionic Bond A chemical bond which arises from the electrostatic attraction between positively an negatively charged ions. Usually a strong bond.

Ionizing Radiation High energy protoplasm-injring radiation that can break covalent chemical bonds or remove electrons from atoms, attaching them to other atoms producing charged ion pairs. This radiation includes u.v., x-rays, y rays and b-particles.

Iron-binding Capacity (IBC) A measure of the percent saturation of transferring in the blood.

Isoagglutin-tion Agglutination of blood cells following a blood transfusion by a different donor blood group. The result is agglutinations of the person's own blood cells by serum of the donor or agglutination of the donor's cells by the person's ownserum.

Isoelectric Focusing Separation of molecules on the basis of the charge. Each molecule will migrate to the point in a pH gradient where it has no net charge.

Isolting Mechanism Geographical physiological, ecological, or behavioural mechanisms that lead to a reduction in the frequency of hybrid matings.

Isomerases Enzymes that catalyze the transfer of two hydrogen atoms from an adjacent carbon to an aldehyde group. This results in the reduction of the aldehyde group to a primary alcohol group, and the formation of keto group on the carbon atom that donated its hydrogen atoms.

Isoniazid Structural analog of pyrioxine; used in the treatment of tuberculosis.

Isotonic Having the same osmotic potential, Said of two solutions. (Contrast with hypertonic, hypotonic.)

Isozymes Chemically different enzymes that catalyze the same reaction.

J

J Gene Gene coding for the joining segment of polypeptide chain which links the *V* (variable regions) to the *C* (constant) regions of both light and heavy chains of immunoglobulins. During lymphoid development the DNA is rearranged so that the *V* genes are linked to the *J* region sequences.

J Segment of Immunoglobulin Genes One of three distinct gene segments encoding immunoglobulin Light chain (the others are V and C) and of four gene segments encoding Heavy Chain (the others are V, D and C). The J (joining) segments encode the last 10 amino acids of the variable region of light and heavy chains. Random genetic rearrangement between light chain V.J. and C, helps to create diversity in antibody specificity.

J-Value The effective concentration of one end of a DNA molecule in the immediate neighbourhood of the other end of the same molecule. J-values and I-values (the total concentration of ends in the reaction mix) influence the formation of products in a litigation reaction. The J-value is a function of the length of a DNA molecule and is independent of the concentration of DNA.

Jerusalem Artichoke A plant (Helian-thus tuberosus) related to the sunflower capable of producing high levels of carbohydrate, in the form of the fructose polymer inulin, in its underground tubers.

Jet Loop Fermenter A type of fermentation system I which agitation and gas absorption are achieved by recirculation of the broth with reinjection into the main fermenter vessel under pressure.

Jordanon A small species, or subspecies, having little variability.

Jumping Gene Transpson that carries genetic information coding forand controlling its own Transposition.

Junctional Complex A site at which there is direct contact between neighbouring animal cells. There are at least four types of special contact between cells; an attchement plaque or desmosome; an adhesie zone; a tight or occuluded junction; and a gap or intermediate junction.

Junk DNA DNA that has no genetic or structural function but persists within the genome since there may be no positive selection against it.

Juvenile The immature, usually non-reproductive [reproductively incompetent], sometimes phenotypically distinct phase of plant growth. Juvenility should nct be confused with young in age although young plants may also be in a juvenile phase. Juvenile phase tissue has repeatedly been found to perform better in culture than adult phase tissue. It is or likely to grow and to form callus, embryoids, shoots etc.

K

Kairomone A transpacific chemical messenger, the adaptive benefit of which falls on the recipient rather than the emitter.

Kallikrein Plasma Proteases normally present as inactive prekallikreins which are activated by Hagemann factor. Act on kininogens to produce kinins.

Kanomycin Antibiotic that operates through its ability to bind to bacterial ribosomes and cause misreading of the message.

Kappa (Paramecium) A particle in the cytoplasm of paramecium that grows in the presence of gene K. producing a killer type.

Kappa Chain One of two types of light chain found in immunoglobulins (for the other,) which differ by about 60 percent in their globulins (for globulins (for the other), which differ by aglobulins (for the other,) which differ by about 60 percent in their constant Regigloulins (for the other,) whiglobulins (for tgloblins (for the other,) which differ by about 60 percent in their constant region amino acid sequglobulins (for the other, JGLGLOBULINS (for the other,) which differ by about 60 percent in their Constant region amino acid sewuences. In anyone immunoglobglobulins (for the other, globulglobulins (for the other,) which differ by about 60 percent inglobulins (globulins (for the globuster. A group of chromosomes radially arranged and star-like in appearance.

Karogamy The fusion of two nuclei that exist within a common cytoplasm, as occurs in the formation of the zygote from two gametes. The process also occurs within the multinucleate plasmodium of fungi belonging to the Myxomycetes.

Karyenchyma Nuclear sapl the morfluid contents of the nucleus.

Karyomites Cromosomes; chromophilic bodies, typically constant in number in any particular species, into which the nucleus resolves itself during cell division.

Karyon A nucleus; the most constant constituent of animal and plant cells which reproduces by mitosis and contains the chromosomes.

Karyoplasm Nucleoplasm; the denser protoplasm of which nuclei are composed as distinct from the less dense cytoplasm outside the nuclear membrane.

Karyosphere Part of the nucleus into which the chromosomes have contracted.

Karyotype The somatic chromosome complement of an individual, usually as defined at mitotic metaphae by morphology (including centromere location and often by special staining techniques, as fluorescence banding) and number, arranged in a sequence that is standard for that organism.

Karyotyping The identification and classification of organisms, cells or tissue on the basis of their chromosome content.

Katachromasis A general term for the changes which take place during telophase and the formation of daughter nuclei.

Kefir A fermented milk product produced in the USSR using a mixture of symbiotic yeasts, lactobacilli and lactic streptococci.

Kelp A number of different genera of large brown seeds, some of which are used as a source of alginci acid, Kelp has also been suggested as a source of biomass.

Keratin Gene since Kerotins are, next to immunoglobulins, the most diverse proteins, the keratin gene family is very large. The genes range in size from 300 to 3000 bases, and some of the smaller ones have no intron. There is a tendency for some keratin genes, and some of these internal repetitious sequences consist of inverted repeats.

Keratins Insoluble protective or structural proteins consisting of parallel polypeptide chains in A-helical or â conformations.

Ketal Boli The heating of worth to inactivate enzymes, precipitate impurities and concentrate the solution prior to inoculation with yeast.

Ketones A collective term for the intermediate products of fatty acid degradation (acetaoacetic acid, beta-hydroxybutyric acid, and acetone). The ketone bodies tend to accumulate when the rate of production exceeds the rate the organism can brun them.

Ketosis The clinical condition in which ketones accumulate in the blood and appear in the urne. This condition is caused by a disturbance in carbohydrate metabolism (uncontrolled diabetes mellitus), dietary diet), or by a reduction in carbohydrate catabolism with resultant high mobilization of body fats (starvation).

Key Factor The hypothetical causative agent most responsible for population density change. In practice, a variable factor in the environment recognized as most closely linked to changes in population density, and measurable by continuous monitoring for use in predicting future population trends.

Killer Yeast Yeast cell containing a double-stranded RNA Plasmid that encodes genes responsible for production of a killer substance which is lethal to other cells. The substance is closely similar in its action and synthesis to a bacterial Colicin.

Kin Selection Component The component of inclusive fitness resulting from helping the survival of relatives containing the same alleles by descent from a common ancestor. (Contrast with individual selection component.)

Kindred A group of individuals each of whom is related in some way, either by blood relationship or marriage, to every other member of the group and, in so far as is or can be known at any given time from the available records, to no other person.

Kinetic complexity Complexity of a sample of DNA as determined by its reassociation kinetic in a renaturation experiment.

Kinetic Constriction Entromere; spindle attachment; insertion region; kinetocheore; a non-staining, localized region in each chromosome which remains single for some time after the rest of the chromosome has divided, and which, at metaphase, appears to be attached to the spindle 'fibre'.

Kenetic Viscosity A description of the flow behaviour of a fluid expressed in terms of the ratio of the dynamic viscosity to the density of the material measured at the same temperature.

Kinetosomes Granular or rod-like bodies often aggregated to form plate like bodies, which occupy the spindly poles in the sporogenetic mitosis of meiosis.

Kinety Row of ciliary Basal Bodies The kineties are composed of linked Kinetetosomes and may be visualized by a silver staining procedure. Kinety pattern is independently inherited, and when artificially altered by surgery is passed on in modified form to progeny.

Kinins A group of inactive peptides in the blood and tissue that, on activation, are involved in blood clotting and inflammatory processes.

Kinosome A minute body sometimes seen in the centromere in such cases there being one kinosome on each chromosome where it passes through the spendle attachment region.

Kirsten Sarcoma Virus Retrovirus originally derived from mice, but subsequently passaged through rats. It contains the oncogene sequence. The human proto-oncogene sequence has not been definitely identified.

Kleinschmidt Spreading Technique Technique developed by A.K. Kleinschmidt in which nucleic acids are spread, in the presence of small amounts of protein, as extremely thin films on cleaned water surfaces. Such spreads may be picked up on copper grids (precoated) and subsequently examined by electron microscopy.

Klenow Fragment Larger part of the bacterial DNA polymerase 1 (76kD) which remains after treatment with subtilisin; retains some but not all exonuclease and polymerase activity.

Koji Process A process by which molds are cultured on the surface of moist bran.

Komberg Enzyme (DNA Polymerase I, pol I) DNA polymerose discovered in 1956 by A. Kornberg and coworkers in cells of E. coli. It catalyses the addition of mononucleotides to the 3'-end of a growing DNA chain provided in vitro with a DNA template, the four deoxyribonculeoside triposphates and Mg^{2+}

Kurtosis The departure of a symmetrical frequency distribution from the normal by excess (platykurtosis) or deficiency (Leptokurtosis in its shoulders as opposed to tails and centre.

Kwashiorkor A nutritional deficiency disease of infants and young children mainly in the tropics and subtropics, occurring soon after weaning. It is caused by a deficient quality and quantity of dietary protein. The disease is characterized by retarded growth, edema, fatty liver, skin changes, severe apathy, and weakness.

L

L Cells Cultured cell line, originally described by Earle in 1943. He took normal subcutaneous connective tissue cells from healthyC3H mice and treated them with 20-methyl-cholanthrene. The L cells thus produced could be cultured and yielded sarcomas when injected into healthy C3H mice.

Laboratory Scale It is a process carried out using only smalls amount (i.e. milligrams or grams) of reagents in order to validate the nature of the products and the characteristics of the reaction.

Lactate Dehydrogenase (LDH) Enxyme that catalyses the inter conversion of lactate and pyruvate in anaerobic glycolysis.

Lactic Acid Bacteria A group of bacteria that fermanent carbohydrates in the presence or absence of oxygen, with lactic acid always a major end product.

Lactoferrin It is iron-binding protein of very high affinity ($K_d 10^{-19}$ at pH 6.4:26-fold greater than that of transferring) found in milk and in the specific granules of neutrophil leucocytes. A protein found in external secretions of vertebrates that binds (chelates) iron.

Lactoperoxidse Peroxidase enzyme from milk which finds an important use in generating active iodine as a nonpermanent radiolabel for membrane protein.

Lactose Carrier Protein The best known example is the product of the lacy gene, coded for in the lactose operon and responsible for the uptake of lactose by *E. coli.*

Lactose A disaccharide carbohydrate (disac. Caride) also known s milk sugar. On hydrolysisit is split into glucose and galactose.

Lag Phase the period of slow, orderly growth when a medium is first inoculated with a culture. Used to describe the first of five growth phases of most batch propagated cell suspension cultures in which inoculated cells in fresh medium adapt to the new environment and prepare to divide.

Lager Beer Beer that has been aged to develop its characteristic taste.

Lagging Strand Strand of DNA that is synthesized discontinuously during DNA replication, and cosists of a series of okazaki fragments lighted together, compare Leading strand.

Lambda Chain One of two types of light chain found in immunoglobulins (for the other) which differ by about 60 percent in their constant region amino acid sequence.

Laminar Air Flow Cabinet or Hood Layered monodirectional airflow. In a laminar airflow (LAF) cabinet air is forced through filters and specially designed ducting to produce an airflow with a minimal lateral disturbance.

Laminarin The chief carbohydrate food reserve of the brown algae. It is a plymer of glucose, is soluble, unlike starch, and closely resembles the callose of higher plant sieve tubes.

Landfil Sit An area of land where domestic solid refuse and similar waste is discharged and buried. Landfill gas is produced in this area. The gas is a mixture of methane (50-65 percent) and carbon dioxide (30-40 percent) together with traces of nitrogen, hydrogen and a wide range of volatile organic compounds generated by anaerobic bacteria present in landfill waste disposal site.

Large Scale Descriptive of a process carried out using the large amounts of material (i.e kilograms or tones) that are similar to those required for commercial production of the product.

Lassa Virus Virulent and highly transmissible arena virus whose normal host is a to dent (Mastomys natalensis); first recorded from Nigeria.

Late Genes Genes that are expressed late in the life cycle of a virus. These genes usually code for proteins such as capsid subunits. Compare Early Genes.

Late-proteins Indicates the proteins synthesized during the later stages of the replication cycle. These are usually the structural proteins necessary for the formation of capside.

Latex It is a complex organic liquid found in certain plants which contains a variety of secondary products, including tremens, terpenoids and alkaloids. Its high molecular weight products founding rubber.

Law of Homologous Series in Variation The principle enunciated by Vavilov. In general, closely allied Linneon species are characterized by similar and parallel series of variations; and as a rule, the nearer these Linneons are genetically, the more precise is the similarity of morphological and physiological variability; generally nearly related Linneons have consequently similar series of hereditary variation.

Layase An enzyme that cataloyzes the addition of groups to double bonds or the formation of double bonds.

Lazy Leucocyte Syndrome A rare and possibly apocryphal human complaint in which neutrophils display poor locomotion towards sites of infection LD (1) Lethal dose. The number of pathogenic microorganisms required to cause death in a given species of animal or plant. (2) lethal dose. Radiation required to kill within a specified time all the individuals in a large group of organisms.

Lead citrate Heavy metal alt used as a stain in the preparation of specimens for transmission electron microscopy.

Leader Peptide A short peptide synthesized in vitro by translation of the leader sequence of some bacterial mRNAs. These peptides are not formed in vivo.

Leader Region A region of regulatory DNA involved in amino acid synthesis, within the promoter-operator region, which is transcribed even in the presence of a repressor.

Leader Sequence Peptide Short length of amino acid sequence at the amino end of some protein molecules, which as a result of hydropoic amino acids, carries the protein through the membrane of the endoplasmic reticulum and so ensures its eventual secretion. The leader sequence peptide is commonly removed before the protein becomes active.

Leading Strand Strand of DNA that is synthesized continuously during DNA replication. The leading strand is synthesized towards the replicating fork in a 5'-to 3'-direction. Compare Lagging Strand.

Leaky Mutant Mutant gene that shows some residual activity.

Lecithin A naturally occurring, chloine-containing phosphiolipid present in animal and plant tissue. Chemically lecithins are similar to fats, but they also contain phosphorus and nitrogen.

Lectins Glycoprotein found on the surface of plant cells. It is cell-agglutinating proteins. Most so far studied have been isolated from plant seeds.

Leghaemoglobin A component of the nodules in nitrogen-fixing plants. It occurs in the bacterio membrane, although it is synthesized on the plant genome. It is haemoglobin-like molecule with a high affinity for oxygen.

Legumin A storage protein (along with vilivin) found in seeds of legumes. The protein has 12 components, each with an average molecular weight of 33,000. It contains more acidic and basic residues than vicilin, and three items as many tryptophan residues.

Leptospira Sprochaete bacterium which causes a mil chronic infection in rats any manydomestic animals. The bacteria are excreted continuously in the urine and contact with infected urine or water can result in infection of humans via cuts or breaks in the skin. Infection causes leptosirosis or Well's disease, a type of jaundice, which is an occupational hazard for sewerage and farm workers.

Lethal (semilethal) Gene A gene that, when expressed before (after) age 1 year, results in the eventual death of the organism. A gene whose phenotypic effect is sufficiently drastic to kill the bearer. Death from different lethal genes may occur at any time from fertilization of the egg to advanced age, Lethal genes may be dominant, incompletely dominant, or recessive.

Lethal Factors Genes which, in the homozygous state, have such a marked deleterious effect that such homozygous organisms are inviable. Where the homozygote is capable of survival but with impaired efficiency the factors responsible are termed Semilethal.

Leucocyte A white blood cell. A number of different types of non-pigmented, nucleate, amoeboid cells which occur in the blood of animals.

Leuconostoc A genus of microaerophilic, spore-forming, Grampoitive cocci that produce lactic acid. Leuconostoc species also produce large amounts of extra cellular dextran. Species of Leuconsostoc are used in the production of dextran and a number of fermented dairy products.

Leukemia An excessive increase in number of leukocytes due to malignancy. It is a form of cancer characterised by extensive proliferation of nonfunctional immature white blood cells.

Leukocidin A substance produced by some pathogenic bacteria and toxic to white blood cells. It kills leukocytes.

Leukotrienes A collection of metabolites of arachidonic acid which have powerful pharmacological effects.

Lichen A mutualistic association of a fungus and either algae or cyanobacteria, resulting in a new structural form. The new formation is morphologically distinct from either of its microbial components and possesses separate chemical and physiological properties.

Life Table A device for expressing in an orderly fashion, observations on the changing density of an insect population in time and space, and the processes which direct those changes especially in relation to the age-specific distribution of mortality and its causes. It is the table showing for a group of equal-aged individuals, the proportion still alive at different times in the future and the number of offspring they produce during each time interval.

Ligament A band of connective tissue linking two bones in a joint.

A tough band or capsule of connective tissue that connects two bones together at a joint. It has a high proportion of elastic fibres and white collagen fibres; therefore it combines strength with elasticity to control movement at the joint. The fibres of ligament penetrate the tissue of the bones, making very strong connection.

Light Chain (l Chain) The smaller of the two types of polypeptide changes (the other is the H chain) that compose the basic Ig molecule; each basic lg molecule contains two identical light chains. Generally refers to the light polypeptide components MW-23,000) of the immunoglobulin.

Likelihood Function The function relating an unknown parameter to observations from which it can be estimated. Cf. *probability function.*

Limited Chromosomes Chromosomes which occur only in nuclei of the germline and never in somatic nuclei.

Lincomycin Antibiotic active against Gram positive bacteria. Acts by blocking protein synthesis by binding to the 50S subunit of the ribosome and blocking peptidyl transferase reactions. Clindamycin, a derivative of lincomycin, is used as an antimalarial drug.

Line Group of homozygous individuals that form a pure breeding population distinct from other individuals of the same species. It is the collection of cells produced by continuously growing a particular cell culture in vitro. Such cell lines will usually contain a number of individual clones.

Linear A term used to describe the physical state of a nucleic acid molecule. The two strands at the end of a linear molecule of double-stranded DNA can be free, as in restriction fragment, bound to a specific protein, as in some eukaryotic extra chromosomes. Eukaryotic chromosomes are linear and their ends, the telomeres, have special sequences which promote their application.

Linearise To introduce a single double-strand break in covalently closed circular DNA molecule and so convert it to a linear molecule. The cleavage of a unique restriction site within a plasmid will linearise it.

Linkage Association between markers on the same chromosome such that they do not show random assortment. Linked markers recombine with one another at frequencies les than 0.5; the closer the markers on the chromosome, the lower the frequency of recombination.

Linkage Analysis The determination of the frequency of crossing-over or recombination of DNA sequence It is used as means of establishing the position of a gene, or sequence, in a chromosome or length of DNA.

Linkage Disequilibrium A condition where two genes are found together in a population at a greater frequency then that predicted simply by the product of their individuals gene frequencies.

Linked Enzyme Assay It is a test procedure in which two or more enzymes are used to catalyze a series of reactions enabling identification and/or qualification of a specific substrate (or levels of a specific enzyme) and the visualization of the result.

Lipid Bilayer An early model for the structure of cell membranes based upon the hydrophobic interactions between phsopholipids. The polar head group faeoutwardlyt the solvent, with the hydrophobic tails clustered in the interior.

Lipid Micelle Amphipathiclipid molecules will not only self-assemble into lipid bilayers in aqueous solutions, but also into miclelles which are small (cololoidal-size), spherical, molecular aggregates in which the hydrophobic tails projects inwards and the surface comprises the hydrophilic poles.

Lipids A large, varied class of water-in-soluble organic molecules. Includes steroids, fatty acids, rostaglandins, terpenes, and waxes. Substance sin a cell which are easily extracted by organic solvents; fats, oils, waxes, steroids, and other large organic molecule, including those which, with proteins, make up the cell membranes.

Lipoic Acid A sulphur-containing fatty acid found in a wide variety of natural materials. It is an essential component in metabolism, although it is active in extremely minute amounts. It has not yet been demonstrated to be required in the diet of higher animals. Lipoic acid is classified with the water-soluble B vitamins.

Liposome Artificial membranous vesicle made from lecithins and other lipids that I readily absorbed by living cell. Liposomes have been exploited for the delivery of novel genetic material to cells as an alternative to mechanical injection.

Lipozygenase Enzyme which catalyzes the oxidative conversion of arachidonic acid to the hydroxyeicosatetraenoic acid (HETE) structure in the synthesis of leucotrienes.

Liqueur An alcoholic beverage produced from aromatic extracts of fruit or other plant material to which sugar is added.

Liquid chromatography A separation technique in which a mixture of compounds in solution is passed through a column packed with a solid matrix.

Liquid Medium Medium not solidified with a gelling agent. Liquid media are used for suspension cultures and for a wide range of research purposes. They are also useful for Stage I cultures in some micropropagation protocols. Usually a support structure or wick is used to hold the tissue above the nutrient medium.

Liquid Scientillation Counter Electronic instrument used to measure quantities of radioactive substances by assay of the emission of radiation during radioactive decay.

Lipmus Paper Red or blue aid-alkali indicator papers. In acids blue litmus paper turns red and in alkalis red litmus paper turns blue. Tehpaper is prepared by soaking absorbent paper in litmus solution and then drying it litmus solution has a pH range from 4.5-8.3. It is obtained from alichen (Leconora tartarea).

Litmus Plant extract, used as a pH indicator, also an oxidation-reduction indicator; it runs blue when alkaline and red when acid in reaction.

Loading The introduction of a feedstock into a reactor or a sample for analysis into an analytical instrument, or onto a gel or similar support material in a chromatographic or electrophoretic technique.

Local Analylaxes A type I hypersensitivity reaction that primarily affects a given organ or tissue (such as skin or nasal mucosa).

Loc and Key Theory The theory, propounded by dufour, that the female and the male genitalia of the same species (atleast ininsects) are so exactly fitted to each other that even slight deviations in the structure of either make copulation physically impossible.

Long Terminal Repeat (LTR) Sequence of several hundred bases fund at either end of some DNA molecules, especially the genomic DNA of retroviruses. The sequences at either end are identical to one another or, in at least some cases, are inverted repeats.

Long-period Interspersion Pattern of distribution of DNA sequences found in the genomes of some organisms such as Drosophila. It involves an alternative distribution of non-repetitive DNA sequences between long stretches of moderately repetitive DNA.

Lost Gene A situation in which all the genes at a certain locus are the same and the other allele has been permanently eliminated from the population.

Lowry Method The most widely used chemical method of protein analysis.

LPR A gene found in MRl mice which is involved in the generation of autoimmune phenomenon.

LPS A produce of some gram negative bacterial cell walls and which can act as a B cell mitogen.

Luciferin A complex compound responsible for emitting light in bioluminescence.

Lucke Virus Herpes virus that causes adenocarcinomas of the kidney in frogs.

Lminol Compound used as a bystander substrate in assaying the metabolic activation of leucocytes by chemiluine scence. When oxidized by myeloperoxidase hydrogen peroxide system, it emits light.

Luria and Delbruck Experiment Experiment in genetics designed to show whether a specific mutation could be environmentally induced or occurred randomly. It involved the use of a fluctuation test, in which numerous cultures of bacteria were simultaneously screened for resistance to lysis by bacteriophage. If mutation to phage resistance were random, then considerable fluctuation in the numbers of mutant bacteria would be expected in the parallel cultures, depending on precisely when the mutation had occurred (if it occurred early in the life of the culture, many progency would have accumulated, it late, no progency would have accumulated). Since marked fluctuation sin the numbers of surviving resistant colonies were obtained, it was concluded tat mutation was occurring at a constant random rate and was not being induced by exposure to the phage or other environmental factors.

Lucury Protein A term sometimes used to describe those proteins which are produced specifically for the function of differentiated cells and are not required for general cell maintenance (the so-called "house-keeping" proteins).

Lymphatic System A closed system of vessel through which the lymph is transported from the interstitial fluid to the blood starch. The flow of lymph in mammals is maintained by muscular activity.

Lyophilize or Freeze Dry To freeze rapidly then dehydrate under high vacuum, the process is Lyopholization.

Lysate Product of cell lysis; often used into refer to a suspension of phage particles released by the lysis of hot bacteria, or to a cell-free homogenate following the action of detergent or mechanical means to break the cells.

Lysis Bursting of a Cell It is a process of disintegration or destruction of host cells during and on the completion of bacteriophage multiplication.

Lytic Growth Developmental phase of a virulent phage during which the viruses multiply vegetatively and kill the host cells lytis at the end of the growth cycle. Lytic growth occurs also following induction of a temperate phage.

Lytic Pathway The sequence of reactions, involved in the expression of the phage genes that results in the completion of a lytic cycle.

The genes of the phage are regulated to permit controlled DNA replication, recombination, and synthesis and assembly of the progency phage. In the prophage stage, all the genes required for forming the lysogemy results. If cro protein is formed progency are formed.

Lytic Viruses Viruses whose multiplication leads to lytis of the host cell.

M

M-ring Innermost (motor) ring of the bacterial flagellar base, located in the outer leaftet of the plasma membrane. It is this ring which is linked to the hook region (and thus to the flagellum itself), and which rotates. Composed of 16 to 17 subunits (one more or less than the S-ring).

Macerase A brand name for peptinase. Pectinase is useful in the isolation of intact cells and protoplasts of higher plants.

Macerate to disintegrate or separate tissues through cutting, soaking enzymatic or other action, resulting in ell dissociation.

Machupo Virus A ember of the renatiridae that may cause a severe haemorrhagic fever in humans. Then natural hosts are rodents and transmission from human to human is not common.

Macro-pyrenic Having nuclei which are considerably larger than the average for the species, individual, etc.

Macroelement One of the main plant nutrient elements essential for plant growth, and required n greater amounts than the others (microelements or micronutrient elements). The microelements are carbon (C), oxygen (), hydrogen (H), nitrogen (N), phosphorus (P), potassium (K), calcium (Ca), magnesium (Mg), and sulphur (S), Si macroelements (N,P, K, Ca, Mg, and S) are added to nutrient media employed forplant tissue culture in the form of salts and are usually required in concentrations above 0.5 mM/ litre.

Macroevolution Evolution above the species level, leading to the formation of genera, families, and other higher taxa; also called trans specific evolution.

Macronucleus (meganucleus) Larger of two distinct sizes of nuclei occurring in ciliated protozoa. These nuclei are highly polypolid and are essentially (absent in the cells of some ciliate strains) have important sexual functions during conjugation. Compare Micronucleus.

Macrotype Egg A type of tachna fly egg which is broadly oval in outline, with a thick, tough dorsal and latral chorion, and a flat, membranous ventral surface by which it is glued externally to the integument of the host: cf, Microtye Egg.

Magnesium An element essential for plant and animal growth. It is contained in the chlorophyll molecule and is thus essential for photosynthesis. In animals it is found in bones and teeth. An magnesium carbonate it is found in large quantities in the skeletons of certain marine organisms, and is founding smaller quantities in the muscles and nerves of higher animals. It is essential cofactor for certain phosphate enzymes, e.g. phosophohydrolase and phosphotrasferase. High concentrations of magnesium ions, Mg2+, are need to maintain ribosome structure.

Magnetic Microcarrier A the of microbead which has magnetic properties and can used in magnetic immunoassay.

Magnetosome Enveloped compartment in magnetotactic bacteria containing magnetite particles. By using this organelle to detect the vertical component of the earth's magnetic field, the bacteria swim towards the bottom of aquatic habitats.

Major Element An element which is essential for life and is required for the formation of the major structural parts of living organisms. They include carbon, nitrogen, oxygen, hydrogen, sulphur, phosphorus, calcium and, in instances, silicon.

Major Histocompatibility Complex A complex of linked genes, with multiple alleles, that control a number of immunological phenomen; it is important in graft rejection.

Malaria Disease caused by the single celled parasite Plasmodium falciparum geographical areas of major incidence of the disease correspond closely with the distribution of sickle cell anaemia

Malate Dehydrogenase An enzyme which catalyzes the reversible reduction of oxaloacetic acid to malic acid in a reaction linked to the oxidation-reduction of NAD(P).

Male Haploidy A system of sex-determination in which the haploid individuals are males whilst the diploid condition determines femaleness.

Male Sterile Maize (genic) A plant producing no viable pollen because homozygous for a recessive pollen lethal usually designated ms/ms.

Malic Acid Fermentation A metabolic process that results in the degradation of malic acid to ethnol. There are two possible pathway based on malic enzyme or malate dehyrogenase.

Malt Extract An extract from dried and ground, germinated grain seed (usually barely, sometimes rice or cornj. It is an undefined constituent of some plant tissue culture media.

Mammalian Cell Cloning The gene manipulation of mammalian cells. One vector that has been widely investigated is the papovaviurs SV-40. This is suitable since its consists of a single small, covalently closed circular DNA molecule, the entire nucleotide sequence of which has been determined. The genomic regions responsible for the viral functions have been established, and information is available on the replication and expression of th evira genome. Tehvirus can replicate eithr vegetatively(in the cytoplasm of the host) or as an integral part of te cellular chromosome. Improved SV40 vectors, which retain the regions implicated in initiation, termination, splicing and polyadenylation, have been constructed.

Manifestation The noticeable effects of a gene. Manifestation may change depending on different external influences or different gene environments, and may sometimes even be entirely lacking.

Mannan A major constituent of the hemicellulose of soft (Coniferous) woods. Its hydrolysis yields mannose, glucose and galactose in the ratio of 3:1:1.

Mannitol [$C_6H_{14}O_6$ mw 182.17] A sugar alcohol widely distributed n plants and often employed as a nutrient and somoticum in plant tissue culture work; as in suspension medium for plant protoplasts.

Mannosidase Enzyme catalyzing hydrolysis of the glycosidic bond between mannose residues and a variety of the hydroxl residues and a variety of the hydroxl containing groups.

Mannuronic Acid A uronic acid structurally related to the hexosemannose. It is a constituent of alginate and other plant gum, as well as occurring in plant cell wall matrix polysaccharides.

Map Distances Distance between two gene loci expressed as Morgan units or some other representation of a map unit.

Map As a verb; to determine the relative positions of genes or restrictions sites on a DNA molecule.

Marasmus An extreme form of chromic under nutrition caused by a lack of calories and protein. Seen mostly in children, it results in loss of subcutaneous ft, inelastic wrinked skin, loss of muscle tissue and strength. It maybe due to chronic malnutrition or defective absorption or utilization of an adequate food supply.

Marker Rescue Phenomenon in which a population of bacteria is infected with two distinct 'marked phages (i.e. two types that can be genetically distinguished), but following the irradiation of one of the two phage strains. Because of recombination, some of the surviving phage come to possess properties of the irradiated strain.

Masked mRNA Messenger RNA present in a cell, but not being actively translated into protein. Masked mRNA is associated with particular types of proteins as messenger ribonucleoprotein (mRNP) particles and can be activated to function in protein sythesis by specific physiological triggers (e.g. hormones).

Mass Selection Artificial selection practiced by choosing in each generation, individuals with maxim (minimum) expression of a given trait as parents of the following generations.

Mast Cell A cell found in the ground substance (matrix) of areolar (loose) connective tissue, which it is thought to secrete. Mast cells are also found in the lining of blood vessels, and some occur in the blood, as basobhils. They have lobed nuclei and cytoplasmic granules that are sustained by basic dyes. Most cells secrete heparin (an anticoagulant) and histamine.

Maternal Chromosomes In bisexual organism the chromosomes derived from the mother.

Maternal mRNA Messenger RNA fund in oocytes and early embrayos which is derived from the maternal characters.

Mating System The pattern of mating in sexually reproducing organism; two types of mating systems are Random mating and Assortative Mating.

Mazam-Gilbert Chemical Sequencing A technique for sequencing nucleic acids based on the differential ability of the four organic bases to react with various reagent such as a piperidine and hydrazine.

Maxicell Cell of the E. coli or other bacteria that has been UV-irradiated, but still retains some genomic DNA. It has multiple copies of plasmids, thus facilitating studies on plasmid transcription.

Maximum Oxygen Transfer Rate It is a measure of the fastest possible rate of uptake of oxygen by a growing culture expressed in terms of the volumetric mass transfer coefficient times the oxygen solubility.

McArdle Disease Hereditary disease in the human resulting from a deficiency of glycogen phoshorylasein skeletal muscle, the gene for which is encoded on chromosome 11.

Mean Corpuscular Hemoglobin (MCH) An expression, in absolute terms, of the average content of hemoglobin of the individual erythrocyte, calculated from the equation:

$$MCH = \frac{\text{Hemoglobin/(gm per 100)} \times 10}{\text{Erythrocyte count } (10^6/\text{mm}^3)}$$

as stated in picograms (10^{-12} gm) per cell.

Mean corpuscular hemoglobin Concentration (MCHC) An expression absolute terms, of the average hemoglobin concentration per unit volume (per 100 ml) of packed erythrocytes, calculated from the equation:

$$\text{MCHC} = \frac{\text{Hemoglobin / (gm per 100 ml)} \times 100}{\text{Hematocrit as stated in grams per 100 ml of packed red cells, or in per cent.}}$$

Mean Corpuscular Volume (MCV) An expression, in absolute terms, of the average volume of the individual erythrocyte, calculated from the equation;

$$\text{MCV} = \frac{\text{Hematocrit (percent)} \times 10}{\text{Erythrocyte count } 10^5/\text{mm}^3 \text{ as stated in cubic microns per cell.}}$$

Medium Formulation One of many available, usually derived formulas that are used in plant tissue culture.

Megalogblastic Anemia Type of anemia caused y a deficiency of folic acid; characterised by an increased level of megaloblasts, the primitive nucleated red blood cells which are much larger mature erythrocytes.

Megasporocyte In plants, the meiocyte that is destined to produce megaspores by destined to produce megaspores by meiosis. Snonym, megaspore Mother Cell.

Meiosis In eukaryoticcell division of a diploid nucleus to produce four haploid daughter cells. The process consists of two successive nuclear divisions with only one cycle of chromosome replication. It is a reduction-division process nuclear division. Daughter nuclei have half the chromosome number of the parent nucleus; occurs in the formation of sex cells.

Meiotic Drive Any alternation of the process of meiosis with the consequence that a parent heterozygous for two genetic alternatives produces an effective gametic pool with an excess of one type (Sandler and Novitskii, 1957).

Melanism Genetically controlled darkening of colour, c Industrial Melanism.

Melanocyte-Stimulating Hormone (Intermedin) A polpeptide hormone produced by the anterior pituitary gland in man. Its role is uncertain although its blood level is often disturbed in patients with endocrine disorders, accompanied by pigmentation abnormalities. It has a marked action on pigmentation in the skin of amphibians.

Melting Curve A graphical plot of the change in absorbance at 260 nm, as a function of temperature that occurs when a solution of DNA is heated. Heating results in denaturation due to breaking of the hydrogen bonds between the bases and dissociation of the two strands of the double helix. The aborbance of double-stranded DNA is always lower tan that of a single-stranded DNA which is lower than that of an equivalent amount of free nucleotides in solution.

Membrane Filter Bioreactor A bioreactor in which the fermenting broth is circulated continuously through a membrane filter. The product diffuses from the reaction zone through the membrane and is withdrawn continuously. This method is also applied to enzyme and resting cell reactors. The hollow fibre fermentor is one variation of this type of bioreactor.

Membrane Transport Transport across biological membranes. Since the lipid bilayer of the membrane is inherently imprermeable to most polar molecules, cells have evolved special ways of transferring such molecules across their membranes.

Mendelian Population A local population of individuals belonging to the same species, in which mating is random.

Mercaptoethanol Strong reducing agent that is often used for the reduction of disulphide linkages in proteins. The reaction converts a cystine which bridges two parts o the polypeptide chain into two chins cystene residues having free thiol groups.

Mercuric Chloride This compound was once commonly used as an antiseptic and fixing agent (0.05-5.10%)for plant material. However, it I highly toxic and is now little used for this purpose.

Meristen Tip A common explant comprised of the meristem

[meristematic dome] and usually one pair of leaf primordial. Also refers to apical [apical meristem tip] or lateral latera or axillary meristem tip] origin. The term meristem tip if often confused with the term short tip, which is much larger and sexually has more immature leaves and some stem tissues.

Meristemoid A cluster of small, isodiametric meristematic cells with a meristerm, or cultured tissue, with the potential for developmental (topipotential) growth. These cells have a dense microvacuolated cytoplasm and a high nucleo-cytoplasmic ration. They may give rise to plant organs [shoots, roots] or entire plantlets in culture.

Meristom Tip A common explant comprised of the meristem (meristematic dome) and usually one pair of leaf primoridia. Also refers to apical (apical meristem tip) or lateral (latral or axillar meristem tip) origin.

Meselson and Radding Model Modificationto the Holiday model of chromosome recombination. A twisting and isomeriztion of strands in the cross-over intermediate is proposed, avoiding the requirements in the original Holliday model for nicking in both strands.

Mesophyll Thin-walled, chloroplast-containing (chlorenchyma), photosynthesizing parenchyma tissue composing the bulk of the leaf between the upper and lower epidermis. The palisade and spongy parenchyma are the two mesophylll cell types of dittos. The mesophyll may also contain collechyma and sclerenchyma cells, which support the veins.

Messenger RNA (mRNA) A kind of ribonucleic acid formed by the copying of certain segments of DNA complementary to it. This type of RNA carries the instruction or the codes, in the form of bases, necessary for the synthesis of a specific protein. It acts as an assembly line or a template in which the amino acids transported by transfer RNAS are arranged in any orderly fashion for the formation of a particular protein.

Metachromatic Stains Stains that colours in cells or tissues which are different from that of the stain solution. For example, aqueous methl violet, differentiated with acetic acid, stains myloid pink/red.

Metaphase Pairing Index The proportion of metaphase cells of which two particular chromosomes have paired, to the total number of cells examined.

Metaphase Plate The group of chromosomes arranged in the equatorial plane of the spindle in the mitotic metaphase and in first metaphase of meiosis.

Metapneustic An insect respiratory (particularly in some dipterous larvae) in which only the last abdominal pair of spiracles is functional: cf. Amphipneustic.

Methanogen A strictly anaerobic bacterium that is capable of producing methane. Methanogens include the genera Methanobaterium. Methanobacillus, Methanococcus and Methanosarcina. Methane bacteria are found in anaerobic soils (marshes and bogs), sewage sludge, landfill sites and anaerobic digestion systems. They also play an important role in the metabolism of ruminants.

Methotrexate Resistance to the drug methotrexate (amethopterin) is a popular selectale marker in cloning experiments with mammalian cells. The drug is aninhibitor of the enzyme dihydrofolate reductase (DHFR) and one route to resistance is the over-production of this enzyme. Thus the inclusion of a DHFR gene in a cloning vector permits the selection of methotrexate transformants.

Metric System Standardized decimal measurement based upon the meter as a unit of length, theram as a unit of weight, and the liter as unit of liquid measure.

MHC (the major hitocompatibility complex) A genetic region founding all mammals whose products are p:imarily responsible for the rapid rejection of grafts between individuals, and which function in signaling between lymphocytes and cells expressing antigen.

MHC Restriction Restriction on interaction between cells of the immune system because of the requirement to recognize foreign antigen in association with MHC antigens (major his to comptibility antigents).

Microbial Pathogen Generally, a microorganis which causes disease in its host; more specifically, a term used in preference to microbial "insecticide" to denote a microorganism used by man to suppress insect pest population.

Microbial Pest Control A technique in which microorganisms (bacteria, fungi or viral culture) or derivates from them, such as the spore protein of bacillus thuringiensis, are used in place of chemical pesticides to kill harmful insects in particular.

Microbial Sensor A biosensor in which the catalyst is a microorganism. For instance, a suspension of anerobic bacteria, such as Pseudomonas fluorescence, is immobilized in a collagen membrane and attached to detect changes in rates of oxygen gas exchange, which in turn are proportional to the glucose concentration of any solution in which it is placed.

Microbiology Study of Microorganism, including the biochemistry genetics, structure, taxonomy, ecology and pathogenesis of microbes, and their applications.

Microbody A small organelle, bounded by a single membrane and possessing a granular interior. Peroxisomes and glyoxysomes are types of microbodies.

Microchromosome In reptiles and especially in bird, the chromosome complement is comprised of large chromosomes and many quite small chromosomes, called microchromosomes. Only recently has it been possible to count their number accurately.

Microelectrode Electrode made of glass, pulled out so that the end tip has a diameter of a fracton of a micrometer, and filled with a conducting electrolyte such as potassium chloride. Such electrodes are used to study the electrical properties of cells ad cell membranes.

Microevolution A small amount of evolutionary change, consisting of minor alterations ingene proportions, chromosome structure, or chromosome number (A large degree of change would be referred to as macroevolution or simply as evolution.).

Microfibrill Threadlike structure fond in the cell walls of filamentous fungi. Chemically, microfibrils consist of chitin.

Micrograft An in vitro graft. This procedure involves placing a meristem tip or shoot tip explant onto a decapitated rootstock, that has been grown aseptically from seed or micropropogated. The purpose may vary, from virus elimination, using meristem tips and virus-free rootstocks, to viral assays or to expedite grafting that is usually done in the greenhouse. The explanted section is placed directly onto the cambium of the severed stock or inserted into a t or inverted T-shaped epidermal incision on the stock. This technique has been used to eliminate viruses, viroids and mycoplasmas from Citrus species.

Microinjection of Cells Procedure whereby molecules (e.g., antibody, mRNA, DNA) or organelles (e.g., nuclei, mitochondria) are introduced into cells by direct injection using ultrafine glass needle. Although relatively large tissue culture cells can be injected. Favoured cells are the eggs of eukaryotic animals, since they are often very large; in some cases direct injection into the cell nucleus is possible. Injection of DNA may lead to permament incorporation of some of the DNA into ıhe nuclear genome of the cell. If the cell is a fertilized egg, this may achieve genetic transformation of the organism and its progeny.

Microperoxidase Part of a cytochome c molecule which retain its haem group and has peroxidase activity. Microphyll A foliage leaf that, even if large, has a singe unbranched vein running from base to apex (some fossil lycopods are an exception). It is the typical leaf of lycopods, horsetails and psilophytes. The stele remains entire when a leaf trace branches off to a leaf. Compare megaphyll.

Microprocessor An electronic device that accepts and processes coded instructions. Microprocessors are now being widely used in the fermentation industry; uses include monitoring routine process variable, data logging, planning maintenance schedules, monitoring inventories to maintain process status information and sending appropriate alarms to operators, and automatic control of fermenter functions.

Micropyle The canal formed by extension of the integuments of the ovule and through which the pollen tube grows into the embryo sac. In the mature seed it appears as a minute pore in the seed coat through which water is imbined when the seed begins to germinate.

Microspore Mother Cell A diploid cell that goes through two divisions, one a reductional, producing four microspores. It is the same as the pollen mother cell.

Microtubule-organizing Centre Region of the cell cytoplasm in which Microtubules are, chiefly synthesized or organized. Many microtubules continue to associate with these centres after growth, although some become linked to other cellulaı structures.

Microtype Egg A minute type of egg produced in large numbers by some Tachinidae, and by the Acroceridae and Trigonalidae. The ventral or poserior end is membranous and used for attachment to a leaf or other surface, whereas the rest of the chorionis thick, tough, and sculptured. In most tachinid microtype eggs, hatching takes place only after ingestion by the host; cf. Macrotype Egg.

Millipore Filter The brand name for a type of membrane filter. These filters come in a range of pore sizes and are used for filtering gases or liquids.

Million's Reagent A mixture of mercury, nitric acid, and water that is used in testing for the presence of proteins. The reagent is heated with the test material and the production of a brick red precipitate indicates the presence of protein.

Minichromosome Chromatin complex of DNA and Histone which is the form adopted by the genome of certain eukaryotic viruses when present in cell nuclei. Viruses such as Papovaviruses replicate in cell nuclei and have double-stranded DNA as their genetic material, ans this viral nucleix cid adopts a nucleosomal organization by combining with histones of the host cell nucleus.

Minimum Effective Cell Density The inoculum densit below which the culture fails to give repreoducible cell growth. The minimum density is a function of the tissue [species, explant cell line] and the culture phase of the inoculum suspension. Minimum density decreases inversely to the aggregate size and division rate of the a stock culture. It is important to know this ratio when plating suspension cultures for selection of single cell lines so that colonies will not overgrow one another prior to obtaining readily manipulable size.

Minor Base It is a nucleotide which differs in structure from the four primary building blocks of nucleic acid (A.C.T.G) and caries a substituent (usually a methyl group), or possesses a saturated bond or an isomeric form of the glycosidic linkage.

Minute Dominant mutation of Drosphila, lethal when homozygous, and thought to involve the coding sequences for Transfer RNA. Affected heterozygous develop slowly; the adults are small, sterile and possess an abnormal pattern of wing veins.

Mitochondrial DNA (mtDNA) DNA of mitochondria. It consists commonly of a closed circle of double-stranded DNA, of between 5 and 3ìm in length, comprising 15-20 kilobases in most animals (longer in higher plants) and including about 50 gene sequences, of which about one-half code for transfer RNA.

Mitochondrial Membranes The two membranes of a mitochondrin differ in structure and composition. The outer membrane has a higher lipid content and a distinct enzyme complement.

Mitochondrial Respiratory Chain A sequences of enzymes and proteins present in the inner mitochondrial membranes of eukaryotic cells and in the cell membranes of prokaryotes. The chain receives electrons or hydrogen atoms derived from reactions of the TCA cycle.

Mitotic Death Cells fatally damaged by ionizing radiation may not die until the next mitosis, at which point the radiation damage to the DNA becomes evident, particularly when there is fragmentation of chromosomes counts.

Mitotic Poison Compound that specifically prevents mitosis, often by blocking spindle formation or operation. Colochincine is a widely used mitotic poison in cytogenetic studies. Some mitotic poisons are used as cytotixic drugs in cancer chemotherapy.

Mixoploid Refers to cells with variable (euploid, aneuploid) chromosome numbers; as mosaic or chimaeral components that differ in chromosome number, or the result of a variety of mitotic irregularities.

Mob The genetic notation for gene(s) involved in the mobilization of a plasmid from cell to cell. Certain broad-host-range cloning vectors are mob to enable them to the mobilized into a large variety of Gram-negative bacteria.

Mobilisation The transfer, between bacteria, of a non-conjugative plasmid by aconjugative plasmid.

Modal Bulk Amodal bulk is produced by determining the mean values for the most important characters, ten discarding all material deviating from any one of these values plus or minus, an amount exceeding the standard deviation and repeating this process several years.

Moderately Repetitive DNA Fraction of DNA obtained by renaturation of DNA fragments from a sample of eukaryotic DNA of reduced molecular weight. Such DNA hybridizes weight. Such DNA hybridizes fairly fast because many copies of the same DNA sequence are present in the solution, but les speedily than does Highly Repetitive DNA or Fold Back DNA.

Modifying Gene A gene which alters the expression of another gene that is non allelic, e.g., genes affecting the amount of spotting in Holstein cattle, a spotted breed.

Modulating Codon DNA triplets (Codons) that code for rare transfer RNA (tRNA) molecules. Since the tRNA which carries the appropriate amino acid to the mRNA is in short supply in the cell, the presence of such codons tends to slow down translation of an mRNA containing them.

Molecular Evolution Evolution viewed at a molecular level. It encompasses the evolution of molecules such as proteins and nucleic acids through the selective competition that occurs in the dynamic biochemistry of the cell as well as the molecular aspects of the evolution of species and organisms resulting from the fact that although selection is at the level of the phenotype, this is itself partially determined by the genotype. Since mechanisms that translate genotype into phenotype are themselves molecular, it is clear that evolution has a very important molecular basis.

Molecular Hybridization Formation of duplexes of complimentary RNA or DNA strands under controlled conditions in vitro, this single-stranded nucleic acids become double-stranded. The progress of the reaction is monitored using methods which can detect this change, such as spectro-photometry or SI nuclease sensitivity.

Molecular Shortcut The differences which distinguish one plant from another are encoded in the plant's genetic material, the DNA. The DNA occurs in pairs of chromosomes (strands of genetic material), one coming from each parent. The genes, which control the plant's characteristics, are specific segments of each chromosome. All of the plant's genes together make up its genome. Some traits, lie flower color, may be controlled by only one gene. But other more complex characteristics, lie crop yield or starch content, may be influenced by many genes. Traditionally, plant breeders have selected plants based on their visible or measurable traits, called the phenotype. But, as indicated by the development of the Selkirk wheat variety, this process can be difficult, slow and costly not only in the development itself, but also for the economy, as farmers suffer crop losses. As a shortcut, plant breeders now use marker-assisted selection. To help identify specific genes, scientists use what are called molecular markers. The markers are a string or sequence of nucleic acid which makes up a segment of DNA. The markers are near the DNA sequence of the desired gene. Since the markers and the genes are close together on the same chromosome, they tend to stay together as each generation of plants is produced. This is called genetic linkage. This linkage helps scientists to predict whether a plant will have a desired gene. If researchers can find the market for the gene. It means the gene itself is present. As scientists learn where each of the markers occurs on a chromosome, and how close it is to a specific gene, they can create a map of the markers and genes on specific chromosomes. These genetic linkage maps show the location of marketers and genes, and show their distance from other known genes. Scientists can produce detailed maps in only one generation of plant breeding. Previously, scientists produced very simple genetic maps using conventional techniques. It was long ago observed that as generations of plants were crossed, some traits consistently appeared together in the new generations (genetic linkage). However, since researchers could concentrate on only a few traits in each attempt at cross-breeding. It took many crosses to obtain even a very simple genetic map. Using very detailed genetic maps and better knowledge of the molecular structure of a plant's DNA, researchers can analyze a tiny bit of tissue from a newly germinated seedling. They don't have to wait for the seedling to grow into a mature plant so that they can test for a specific characteristic. Once the tissue is analysed, scientists know whether that seedling contains the appropriate gene.

Molecular Weight Marker A protein, polypeptide, nucleic acid, oligoucleotide, etc. used as a reference material in analytical techniques such as gel filtration chromatography, electrophoresis and density gradient centrifugation.

Moloney Murine Leukaemia Virus Replication competent retrovirus (Oncovirinae) which causes leukaemia in mice, isolated by Moloney from cell-free extracts made from a transplantable mouse sarcoma.

Monocional Antibody An antibody preparation which contains only a single type of antibody molecule. Monoclonal antibodies are produced naturally by myeloma cells. Amelomais a tumour of the immune system. A clone of cells producing any single antibody type may be prepared by fusing normal lymphocye cells with myeloma cells to produce a hybridoma.

Monoclonal Antibody Antibody synthesized and secreted by individual clonal populations of hybrid cells-hybridomas-produced by the fusion of individual a lymphocytes from an immunized animal with individual cells from a lymphocytic tumour. The tumour cell crenders the hybridoma cell immortal, and the b cell continues to secrete its single specific antibody species. Hybridomas comprising each selected clone can thus be propagated and grown in bulk to obtain, from each one, copious amounts of specific and homogeneous antibody.

Monoculture The cultivation of a single crop species over large areas without provision for diversity or use of the land in any other way.

Monoecious Organisms in which both sexes are "housed" in single individual, which produces both eggs and sperm. (in some plants these are found in different flowers within the same plant). Examples; corn, peas, earthworms, hydras.

Monogenic or Monohybrid Ration The 3:1 ratio obtained in the F_2 of a cross between two parents of a particular strain differing in respect of only one pair of contrasting characters.

Monohybrid Heterosis Hybrid vigour due to heterozygosity at a single gene locus.

Monopodial Growth A type of growth in pants, characterized by a branching system in which the primary apex continues to grow and remains dominant. Morbilli virus Genus of viruses (of the Paramyxovindae). Type species is measles virus; other species include canine distemper virus and the related seal virus.

Morphactin One of a number of structurally diverse compounds that possess plant growth activity and react synergistically or antagonistically with endogenous plant growth substances, possibly as inhibitors of auxin transport.

Morphogen Substance that induces specific changes in morphogenesis. The word is mot frequently applied to a substance that acts like a hormone, in that the target site or tissue is other than or distant from the site of synthesis. The trims probably sunonymous with inductor.

Morphosis Non adaptive, usually unstable variation in Morphogenesis, Associated with certain environmental changes. These phenocopies may mimic the effects of known mutations.

Mortality, Fatality The ratio of the number of deaths to a given population in a defined situation.

Mother Cell A cell with a diploid nucleus which by melosis, gives fur haploid nuclei, e.g. the spore mother cell in Bryophyta and Pteridiphyta, the Microspore or Pollen mother-cell and the Megaspore or embryo-sac mother cell in Flowering Plants. In animals these perm mother-cell is known as the spermatocyte and the egg mother-cell as the oocyte.

MPN Most probable number, a statistical expression of estimating cell number in a culture.

Multi or Polygenes Two or more pairs of genes contributing to the expression of one and the same quantitative character such as yield, size, height intelligence. In other words, each gene has an additive effect and the character expressed is dependent on the cumulative effect of the gene present.

Multi-enzyme System A group of enzyme active in the sequential steps of a metabolic pathway and in physical proximity to one another.

Multigene Family Groups of genes which are similar in fine structure, have overlapping functions, and are closely linked.

Multimeric Refers to protein units formed out of several monomers and themselves comprising the capsomeres.

Multiple Factor hypothesis hypothesis A hypothesis which explains the blending inheritance often in quantitative characters ion the assumption that there is a series of independent cumulative genes governing such a quantitative trait.

Multiplication or Proliferation (1) Any tissue culture undergoing rapid growth. (2) Commonly used to describe Stage II culture, in which consecutive subcultures promote an increase in the number of propagules. (3) The media employed to induce culture proliferation (multiplication media).

Multiplication Rule If an event occurs in n_1 ways, and a totally independent even can occur in n_2 ways, the number of ways both events can occur at the same time is n_2 x n_2.

Muscle Proteins Those proteins present within muscle fiber which are involved in the contractile process Includes action, myosin, tropomysion, and –actinin.

Mutagens Physical or chemical agents, such as radiation, heat, or alkylating or dominating agents, which raise the frequency of mutation greatly above the spontaneous background level.

Mutation Rate The frequency with which mutations take place in a given variety or species, or the frequency with which any specified mutation occurs in a given population.

Mutation Spectrum A term indicate that on organism may produce different kinds of mutants due to the differential action of various mutagens.

Mutation Trend A series of slight gene mutations in the same direction resulting in gradual intensification of a character change.

Structural Mutation Intergenic change in the linear arrangement of genes, sometimes producing a Position Effect.

Mutation Rate The proportion of individuals or cells in a given group which show mutation for a given gene under given conditions in one generation, or other staged unit of time.

Mutational Load Accumulation of deleterious genes resulting from Mutations genetic disadvantage sustained by a population of organisms as a result of such mutations.

Mutator Gene A gene that increases the mutation rate of other genes in the same organism.

Mutualism The type of symbiosis, such that exhibited by fungi and algae or cyanobacteria in forming lichens. In which both species profit form the association.

Mycellum One of the mass of interwoven threadlike filaments forming the vegetative portion of the thallus in fungi.

Mycobacterium Microti Mycobacterium which causes tuberculosis-like disease in small rodents (Microtus microlus is the vole); will infect mice but not humans, and is therefore much used a laboratory model. Releases large amounts of cyclic AMP which may inhibit lysosome- phagosome fusion.

Mycoplasma They are prokaryote, without cell wall and have been placed under the class mollicutes (l. Mollis = soft, pliable + Cutis = stain) an the order mycoplasmatales. They are smallest micro-organism which have been known to cause a number of diseases in animals and human beings. They are characterised by their lack of cell wall, the outer boundry of the cells being the cytoplsmic membrane. Therefore, their cells possess plasticity and can assume various shapes ranging froj spheres to branched filaments. The plasticity allows the cells to pass through bacteriological filters even though the smallest cells are about 0.3 nm in diameter. They are unicellular, invariably non-motile (except Spiroplasma), prokaryotic mini-lives (organism) incapable of synthesizing cell wall or cell wall-materials.

Mycotoxicosis Poisoning of humans and animals following the ingestion of foods containing toxins; produced by fungi.

Myelin A material forming a sheath around same axons. It is formed by Schwann cells that warp themselves about the axon. It serves to insultate the axon electrically and to increase the rate of transmission of a nervous impulse.

Myoglobin A conjugated protein found in muscles (sometimes referred to as muscle haemoglobin'} It is similar to haemoglobinin being a haemoprotein capable or binding oxygen but is structurally simpler, having only one polypeptide chain combined with the haem group. Each molecule of myoglobin can attach one atom of oxygen.

Myosin Protein molecules, each composed of two coiled subunits (MW-220,000), which can aggregate to forma thick filament which is globular at each end.

Myxobacteria Saprophytic gliding bacteria that may produce brightly coloured fruting bodies visible to the naked eye. The cells, which occur in masses of slime, are flexible unicellular rods that divide by transverse fission. Myxobacteria are common in soil, animal dung, and decaying plant matter.

N

1-Naphathaleneacetamide A synthetic hormone analog of the auxin type. It is sometimes used in plant tissue culture media. It dissolves in base.

2-naphthaleneacetamide Acetic Acid or 2-naphthoxyacetic Acid or b-naphthoxyacetic Acid [NOA or b NOA, $C_{12}H_{10}O_3$ mw 202-20] A synthetic hormone analog of the auxin type. It is sometimes used in plant tissue culture media and in horticulture to promote rooting of cuttings. It dissolves in base [ca. 1M HOH or NaOH].

N-terminus The amino (-NH_2) end f peptide chain, by convention written as the left end of the structural formula.

NAD (Nicotinamide Adenine dinucleotide) A compound found in living cells, existing in two interconvertible forms; the oxidizing agent NAD and the reducing agent NADH.

Naegleria Genus of amoebae found in the solid and best known for their transition from an amoeboid to flagellated cell form in experimental conditions involving changes in ionic content of the solution in which the cells are suspended.

Nagative Control The situation in which a regulatory macromolecule (generally a repressor) functions to turn off transcription. In the absence of a regulatory macromolecule, the structural genes are turned on.

Nail Patella syndrome in herited disease in the human, due to a dominant gene on chromosome 9, and expressed in the phenotypeas small or absent kneecaps and deformed fingernails.

Naphthaleneacetic Acid A growth regulating substance which has been used with considerable effect, at low concentrations, in the induction of flowering.

Nascent Describing a molecule that is in the process of synthesis and therefore still incomplete. The term is applied chiefly to the growing chains of polypetptides or polynucleotides on ribosomes or on the DNA.

Native Immunity A state of resistance conferred by one's genetic make-up, not acquired by exposure to infectious disease agents.

Natural Complex A complicated, non-synthetic, often undefined addendum in plant tissue culture media, such as organge or tomato juice or coconut milk.

Natural Control The process of dynamic equilibrium which maintains the characteristic mean density of wild population within particular upper and lower limits, over a period of time, by a complex combination of all the additive, conditioning and subtractive processes impinging upon that wild population.

Nature and Nurture The antithesis between Genotype and Environment.

Nearest Neighbour Analysis A technique used in the structural analysis of oligonucleotides. It is based on the fact that when a new nucleotide is added to the growing point of a transcript the A-phosphate on its 5'-hydroxyl is linked to the 3'-hydroxyl of the preceding nucleotide.

Nef (Nephritic Factor) An autoantibody to the alternative pathway C3 convertase which stabilizes the enzyme and potentiates the action of the alternative pathway amplification.

Negative Complementation Situation when complementation between alleles permits a mutant subunit to inactivate the wild-type subunit with which it associates in a mulimetric protein.

NEGATIVE Stain Stain which colors the background while the cells appear clear, or unstained.

Negative training Staining technique used in the electron microscopy of viruses and other particulate material. S suspension of the material to be examined is mixed with a solution of phoshoungstic acid ad sprayed onto a carbon-coated grid by an atomizer. The fine detail of the contours of the specimen is rendered more evident by this technique.

Neonatal Mortality Rate The number of deaths reported among infants under one month of age in a calendar year per 1000 live births in the same year and place.

Neoplasm Localized cell multiplication or tumor; a collection of cells which have undergone genetic transformation. These cells differ in structure and function from the original cell type.

Neoplastic Growth Indicates tumorous growth arising out of abnormal localized mltiplication of certain types of cells; malingnant, if the growing cells infiterate surrounding tissue and/or are carried by blood or lymph to other locations in the body, growing there more vigorously. Otherwise benign i.e., localized at a particular place only and relatively less harmful.

Neotency The retention of larval or other juvenile features beyond the normal stage in the development of an animal. It may be either temporary, because of climatic or other factors, or permanent, in which case the animal breeds in the larval stage. Neoteny is thought to be important in the evolution of some groups including man, who has certain resemblances to young stages of apes.

Nerurofilament Filment within a nerve cell, composed of three different intermediate filament polypeptides and though to have a role in strengthening axons.

Nerve Cord An enclosed cylindrical tract of nerve fibres that forms a central route for the conduction of nerve impulses within the body. Vertebrates and other chordates have a single hollow nerve cord (the spinal cord) situated dorsally. Invertebrates generally have two or more nerve cords, each lacking a central cavity and wit ganglia situated at intervals along its length.

Neural Cell Adhesion Molecule (NCAM) Cell surface glycoprotein of vertebrates initially defined by antiserum capable of inhibiting cell adhesion of chicken neural retina cells. Believed to be important in calcium-in-dependent intercellular adhesion of neutral and other cells, including cells in early embryos.

Neuraminidase Enzyme catalyzing cleavae of neuraminic acid residues from oligosaccharide chains of glycoproteins and glycolipids. Since these residues are usually terminal, neuraminidases are generally exoenxymes, although an endoneuraminidase is known. For use as laboratory reagent, common sources are from bacteria such as Vibrio or clostridium. A neuraminidase is one of the transmembrane protein of the envelops of influenza virus.

Neuroblast Any of the cells of animal embryos that become or produce nerve cells, frequently by unequal mitotic divisions. Although many of the cells of the neural tube of vertebrates will produce nerve cells, the term neuroblast is usually restricted to those cells whose immediate progeny will differentiate as nerve cells. Neuroendocrine Systems · The systems involving both nervous and endocrine factors that control functions of he body, Many examples are known, particularly involving the pituitary gland, from which hormones are secreted under direct nervous stimulation.

Neurohormone A hormone that is produced by specialized nervous tissue. Examples are noradrenaline, serotonin, vasopressin, and oxytocin.

Neurokinin Biologically active peptide; one of a class of peptides known as tachykinins.

Neurological Mutnt Organism displaying abnormalities of locomotion or behaviour traceable to malformation or malfunction of the central nervous system.

Neurone Theory The theory, now accepted, that the nervous system is composed of many discrete nerve cells (neutones), closely contacting each other at synapses. It replaces the reticular hypothesis that there is cytoplasmic continuity between cells.

Neurotens in tridecapeptide hormone (sequence : elyenkprpyll) of gastrointestinal tract: has general vascular and neuroendocrine actions.

Neuter Describing individuals that lack sex organs. For example, some flowers in the family Compositae lack stamens and pistils.

Neutral Mutation Mutation that confers neither selective advantage or disadvantage on the organism concerned. According to a theory advanced by M. Kimura, most nucleotide substitutions results in neutral or near-neutral mutations. Such mutations may spread in a population by chance since there is no selective pressure against them. However, many geneticists disagree with Kimura's interpretation and hold that neutral mutations are rate or non-existent, the argument turning on how fine an advantage or disadvantage must be to be significant or measurable.

Neutral Red Biological stain and acid/base indicator which gives a red colour in acid; and an orange-brown in basic conditions. It is normally employed as a solution in 70 percent ethanol.

Neutral substitution Amino acid substitution in a protein that does not affect the activity of the molecule.

New Reunion (of chromosomes) Rejoining between a broken end and any other broken and except the one to which it was originally attached.

Newcommer's Fluid A fixative used in chromosome analysis composed of 6 parts isopropyl alcohol; 3 parts propionic acid: 1 past petroleum ether : 1 part aetons: 1 part dioxane (with or without ferric acetate).

Niacin Test A diagnostic test used to detect those species of Mycobacterium which produce-and liberate into an agar medium an easily detectable amount of niacin (i.e. nicotinic acid).

Nick Translation A means of introducing radioactively labelled nucleotide triphosphates into native DNA in vitro, employing *E. coli* DNA polymerae 1. Using a 5'-PO_4/3'-OH single-strand nick in a native DNA molecule, DNA polymerase 1 catalyzes the addition of an exogenous radioactive nucleotide to the 3'-OH side of the nick and hydrolyzes a nucleotide from the 5'-PO_4 side of the nick. This moves the nick along the native DNA molecule by one nucleotide. Subsequent polymerization and concomitant hydrolysis translates the nick along the DNA molecule while introducing more radioactive nucleotides into the DNA.

Nicking-closing Enzyme (topoisomerase 1) enzyme involved in the normal DNA replication. The enzyme in *E. coli* is a monomer (Mr = 110 000) which uses negatively super coiled DNA as a template. Its function is to relax super coiling stress which would other wise build up in the DNA during unwinding for replication.

Nif The genetic notation for the genes involved in nitrogen fixation. The *nif* operon is a complex array of a seventeen genes. The proteins encoded by the *nif* genes will fix atmospheric nitrogen (N_2) into ammonia (NH_4^+) and nitrate NO_3). Many soil bactria will fix nitrogen and there is much interest in manipulation of *nif* genes from bacteria to allow plants to fix nitrogen.

Ninhydria A reagent used to test for the presence of proteins andamino acids. An aqueous solution turns blue in the presence of alpha amino acids in solution. When dissolved in an organic solvent it is used as a developer to colour amino acids on chromatograms. If a chromatogram treated with ninhydrin is heated strongly the amino acids appear as purple spots that can be identified by measuring the Rf value. Ninhydrin is carcinogenic.

Nitrate Reductase An enzyme that catalyzes the conversion of nitrate to nitrite. Eukaryotic nitrte reductase is an enzyme complex containing flaving (FAD), haem (cytochrome b557) and molybdenum. Electrons are provided by NAD(P)H derived from photo-synthesis or carbohydrate oxidation. Prokaryotic nitrate reductase is a smaller protein containing molybdenum, but no haem or cytochrome. The enzyme has a fast rate of turnover and is present at high concentrations when nitrate is available, but is repressed by the presence of ammonium ions.

Nitrite Reductase An enzyme that converts nitrite to ammonia. Plant and photosynthetic prokaryotes possess a ferredoxin-linked nitrite reductase which accepts electrons from reduced ferredoxin, but not from NAD(P)H. the enzyme contains a tetra nuclear iron sulphur centre and a special haem (siro similar to that founding sulphite reductase. Nitrite is reduced directly to ammonia without the formation of free intermediates.

Nitrogen Balance Measurement of nitrogen equilibrium of the body.

Positive balance occurs when new body protein is being formed (pregnancy, growth and healing); negative balance occurs when body protein is being broken down (fever, surgery, burns).

Nitrogen Fixation Conversion of nitrogen gas to ammonia, which makes nitrogen available to living things. Carried out by certain prokaryotes, some of them free-living and others living within plant roots.

Nitrogenase An enzyme complex that catalyses the reduction of atmosperic nitrogen (dinitrogen) to ammonia in nitrogen-fixing organism. The enzyme has two major components, each of which ahs two or four subunits. The large component contains molybdenum, non-haem iron and sulphur. The enzyme, which is strictly anaerobic and is easily inactivated by molecular oxygen, also acts as a hydrogenase. Hydrogen production is a wasteful reaction, and those species with a low level of hydrogen production are more efficient nitrogen fixers.

Nitrogenous Base An aromatic N-containing molecule having basic properties (tendency to acquire an H atom). Important nitrogenous bases in cell are the purines and pyrimidines.

NK (Natural Killer) Cell Natural killer cells are group of lymphocytes which have the intrinsic ability to recognize and destroy some virally infected cells and some tumour cells.

Nodal Culture The culture of nodal explants, usually including a lateral bud and a section of adjacent stem tissue.

Nomogenesis The theory that the evolution of organisms is the result of certain processes inherent in them and that it follows definite laws.

Non-coding DNA DNA that is in the nonsense strand (negative strand) of the DNA duplex, or is out with the coding sequence of any gene. It may or may not be taken to include intron sequences. Centrometric region DNA is non-coding.

Non-equivalence Term in cell determination for cells which will give rise to the same sorts of differentiated tissues but which have different positional values (EG. cells of fore-limb and hind-limb buds).

Non-permissive Condition Condition that prevents survivors of conditional lethal mutations.

Non-reciprocal Contact Inhibition Collision behaviour between different cells types in which one cell shows contact inhibition of locomotion, and the other does not. An example is the interaction between sarcoma cells and fibroblasts (the former not being inhibited).

Non-repetitive DNA DNA that shows reassociation kinetics indicative of unique DNA.

Non-transcribed Spacer DNA sequence that lies as a spacer between tandem copies of a transcribed gene sequence, but is not itself transcribed. Note that the so-called non-transcribed spacer, between the copies of the large ribosomal genes in Xenopus is now known to be transcribed.

Nonrandom Mating A mating system in which the frequencies of the various kinds of matings with respect to some trait or traits are different from those expected according to chance.

Nopaline One of the opine compounds that are synthesized by plant cells when present in crown gall and stimulate the agrobocteria causing the gall to grow rapidly.

Normal Curve A bell-shaped curve which represents the distribution of individuals segregating for a quantitative character, such as, height, yield intelligence etc. usually the curve is humped in the midway between two extremes indicating that the majority of individuals in the population are groped in this class called mode or model class. There may be two modes. The mode may be nearer tone end of the curve in which case it is called a skewed curve.

Nucellar Embryo the process by which individual cells of the nucellus give rise to somatic embryos [embryoids]. Plants on which this occurs naturally are considered good candidates for tissue culture of somatic embryolds.

Nucellar Embryony A form of apomixes where the embryo arises directly from the nuclellus.

Nuclear Budding The division of a nucleus into daughter nuclei by constriction

Nuclear Disc The star-like group of chromosomes on the equatorial plate at metaphase.

Nuclear Magnetic Responance Spectroscopy [NMR spectrosocopy). Technique used in chemistry and biochemistry to identify different compounds in complex mixtures and to quantify the amounts present by analyzing absorption specta following changes in the resonance frequencies of the atomic nuclei.

Nuclear Pore An annular-shapped hole that occurs in the nuclear membrane; usually large numbers are present. The inner and outer membranes of the nucleus become fused together at the annuleus. Nuclear pores are the site of active transport of materials into and out off the nucleus. Nuclear imports include the nucleotide precursors of DNA and RNA, histones and ribosomal proteins, whereas exports include mRNA, tRNA and ribosomal subunits.

Nuclear RNA The nucleus contains RNA which has just been synthesized, but in addition there is some which seems not to be released, orisonly released after further processing, the heterogenous nuclear RNA (hnRNA) and small RNA molecules associated with protein to form small nuclear ribonucleoproteins (snRNP).

Nuclear Transplantation Experimental approach in study of nucleo-cytoplasmic interactions, in which a nucleus is transferred from one cell to the cytoplasm (which may be anucleate) of a second.

Nucleases A kind of hydrolytic enzyme (hydrolase), capable of hydrolyzing nucleic acids into their constituent nucleotides, are of two types, amely, DNase (deoxyribonuclease) and RNase (rbonuclease) hydrolysing DNA and RNA respectively.

Nucleic Acid Sequence Analysis A technique for determining the sequence of bases in a specific nucleic acid and from which the amino acid sequence of the protein for which it codes can be determined. It is also used in determination of taxonomic relationships between viruses and other organisms, as well as determining control sequences governing initiation and termination of transcription. Nucleic Acids Compounds of pentose sugars, phosphoric acid and nitrogen containing base purine and pyrimidine); these three units combine to form nucleotides which in their turn join to form polynucleoide strands generally referred to as nucleic acids. These are of two types, ribonucleic acid (RNA) and deoxyribonucleic acid (DNA); the former has ribose as the sugar and is generally single stranded, the latter has deoxyribose as the sugar and is generally double stranded. RNA is mostly found associated with cytoplasm and DNA with nucleus in a cell. Both the types are characteristic of cells of all living organisms. However, in viruses only, either of the types is observed DNA in considered to be the universal genetical material and is the repository of the genetic code. RNA plays an important role in translating this code in terms of proteins, particularly enzymes.

Nucleocapsid The fundamental nucleic acid of a virus, plus its surrounding "crystalling" protein coat or shell. Some nucleocapside are naked, but a mature, ultimate phase virus particle is enclosed by a limiting membrane, and is known as a ririon.

Nucleolar Organizer A region on a chromosome that is associated with the formation of a new nucleolus following nuclear division. The site of the genes that code for ribosomal RNA.

Nucleosome Phasing The arrangement of the nucleosomes with respect to specific DNA sequences within a eukaryotic chromosome. Nucleosome phasing is thought to control the expression of some eukaryoticgenes. If the DNA sequence containing a gene's promoter is buried within a nucleosome it is inaccessible to RNA polymerase and the gene is not expressed. On the other hand, if he gene's promoter is contained in an internucleo some space, them RNA polymerase may bind to the promoter and express the gene. Changes in nucleosome phasing have been correlated with developmental switches in some eukaryotic organisms.

Nuclide Atom designated according to the constitution of its nucleus in terms of proton and neutronnumbers. Thus radioisotopes may be described as radionuclides.

Null Hypothesis The hypothesis from which the expectations are formulated for the purpose of a test of significance.

Nulliplex Condition of a polyploid in which all the chromosomes of one homologous type carry the recessive allelomorph of a particular gene, as opposed to simlex, duplex, triplex, quadruple, etc, in which the dominant allelomorph is represented one, two, three, four, etc., time (Belling, Blakesless and Farnham, 1923)

Nurse Cell Cell in the ovary of insects that provides nutrition for the developing oocyte. The cytoplasm of the nurse cells streams into the occytes, which therefore grow at their expense. The cycles of DNA replication in nurse cells are endomitotic (i.e. no nuclear division occurs).

Nutrient Agar A nutrient both solidified by the addition of agar.

Nutrient Film Technique (NFT) Hydroponic plant growth whereby plant roots are suspended in shallow, slowly circulating nutrient solutions delivered as a continuous film of liquid rather than in bursts of liquid.

Nutrient Limitation The condition in which growth of a culture is limited by the availability of a nutritional requirement. The lack of a carbon source or any of the essential elements, vitamins or other compounds leads to a cessation of growth. Lack of other nutrients may produce change sin metabolism.

O

O Antigen Antigen of cell walls of enterobacteria such as E. coli and salmonella.

Oat Coleoptile Test A quantitative auxin assay. This determination is based on the degree of curvature, a result of asymmetric growth, in a detipped oa coleoptile affixed to agar containing the auxin.

Obelin Calcium-activated photoprotein in the photycyte of the colonial hydroid coelenterate, obelia geniculata.

Obligate Parasite One that is unable to multiply a part from a living host.

Obligate Pathogen A disease-causing microorganism which requires a living host in which to grow and reproduce: cf. Facultive Pathogen.

Occam's Razor Scientific principle that the simplest adequate hypothesisis preferred: named after the scholar William of Occam in 1347. Thus, when several possible explanations of a phenomenon are offered, the one favoured involves the fewest assumptions and is most consistent with the available data. It is also known as the parsimony principle.

Occult Virus (Latin occultus, hidden) Refers to a hidden virus i.e., a virus causing in apparent infection (q.v.) and thus remaining latent (q.v.)

Ochre Codon The codon UAA (in mRNA) which odes for polypeptide chain termination.

Ochre Mutation Nonsense mutation that changes the nucleotide sequence of a gene to give the codon UAA. This results in premature termination of polypetide biosynthesis.

Octopine Plant growth substance-an opine-produced by cells in crown galls in plants and active in promoting growth of the Agrobacterium in the crown gall cells.

Odontoblast A cell that secretes the dentine of a tooth. Odontblasts lie around the outside of the pulp cavity and their long fine processes extend into fine canals 9canaliculi) in the dentine.

Oestriol A metabolite of oestradiol and oestrone, found in he urine in high concentrations. It has only weak oestrogenic activity.

Oestrone A metabolite of oestradiol, found in the urine in high concentrations. If has onlyweak oestrogenic activity.

Oestrous Cycle The reproductive cycle that occurs in sexually mature, non-pregnant female mammals. It is regulated by the hypothealamus which release factors that stimulate the periodic secretion of pituitary gonadotrophins.

Offset A short runner growing horizontally above ground away from the parent, seen in daisies and houses-leeks. It stores no food but turns up at the end and produces a new plant from the apical bud. Like stolons and suckers, offsets area means of vegetative propagation in angiosperms.

Oil Immersion Amicroscopic technique using special high-powered objective lenses. A drop of immersion oil (e.g. cedarwood oil) is placed on the coverslip of a microscope slide and the objective lens carefully lowered into it. The oil has the same refractive index as the lens glass and increases the resolving power obtainable by letting a wider angle of rays enter the objective lens.

Okazaki Fragment Segments of approximately 1,000 to 2,000 nucleotides of the daughter strand of replicating DNA. These are joined in succession by the enzyme DNA ligase to form a normal double-stranded DNA molecule.

OKT A group of monoclonal antibodies used to identify T cell surface markers in humans.

Oleic Acid Descriptive of a cell that accumulates large amounts (over 20 percent of the dry weight) as lipids.

Olfactory Nerve (Cranial nerve 1) the nerve that connects the olfactory cells of the nose with the olfactory bulb at the front of the vertebras. The cell bodies of its sensory fibers are located not in a ganglion but in the olfactory membrane.

Olfactory Organ An organ of smell. In vertebrates, olfactory organs are paired structures located in nasal cavities that communicate with the exterior environment through the external nares. In vertebrates, olfactory organs includes structures such as the antennae of insects.

Oligo (dT) Cellulose Short lengths of deoxythymedilic acid (dT) cellulose is used in the purification of eukaryotic mRNA by affinity chromatography. In this process the poly(a)-tail of the mRNA molecule is immobilized by hybridization with the oligo (dT)

Oligophagous (stenophagous) Adapted to the use of only a limited number of plant or animal species (e.g. members of a genus) as hosts or prey: cf. Monophagous, Polyphagus.

Oligotrophic Describing lakes that are deficient in nutrients and consequently low in productivity. Compare Eutrophic.

Ommatidium [Gr. Omma: an eye]: One of the units which, collected into groups of up to 20,000 make up the compound eye of arthropods.

On-line Descriptive of the continual transfer of data from a recording device to a data logger, chart recorder or microprocessor.

Ommatidium One of the units which, collected into groups of up to 20,000 make up the compound eye of arthropods.

Omnivore An organism that can eat both animal and plant material. (contrast with carnivore, deteritivorce, herbivore.)

Omnivore An animal that feeds on animal and plant material. On-line Descriptive of the continual transfer of data from a recording device to a data logger, chart recorder or micoporocessor.

On-line Descriptive of the continual transfer of data from a recording device to a data logger, cart recorder or microprocessor.

Oncogene Theory A theory to explain the viral etiology of cancer. It proposes that the cells of all vertebrate species contain, incorporated n their chromosomes, a potentially carcinogeni provirus, called an oncogene, and that oncogenes are kept in a repressed state in the normal cell by repressor proteins. The transformation of a normal cell to a cancer cell occurs when the repressor protein specific for the oncogene is blocked.

Oncogenes Genes that have oncogenic or cancer-causing potential. They are initially discovered in retroviruses, although they are not native to these viruses. The protein products of many of the oncogenes studied are protein kinases or protein phoshory lating enzymes.

Oncogenic Virus A virus that carries an oncogene which may be inserted into the genetic elements of an infected host cell or organism in abnormal development. Both RNA and DNA viruses may be oncogenic. They include adenoviruses, leukoviruses and papovaviruses. Oncogenic viruses are also capable of inducing cell transformation *in vitro*.

Oncovirinae The family of retroviruses (Retroviridae) which can cause tumours. They are enveloped by membrane derived from the plasma membrane of the host cell, from which they are released by budding without lysing the cell. Within each virion is a pair of single-stranded RNA molecules Replication involves a DNA intermediate made on an RNA template by the enzyme reverse transcriptase.

Ontogeny The development of a single organism in the course of its life history.

Oocyte Primary egg mother cell giving rise by the first meiotic division to the secondary oocyte and the first polar body. The secondary occyte at the second meiotic division gives rise to the ovum and to a second polar body.

Oogenesis The process in sexual organisms, particularly in aminals and in man, by which female gametes (egg) are formed.

Oosphere A female gamete or egg-cell. In angiosperms it is enclosed in the embryo sac and protected by the integuments of the ovule. In certain fungi (e.g. Pythium) the oosphere is thin-walled but, after being fertilized, develops a thick resistant wall and becomes an oospore.

Opal Suppressor A gene which codes for an altered transfer RNA so that its anticodon can recognize the opal codon and thus allows the continuation of protein synthesis. A suppressor of an opal mutation is a tRNA which is charged with the amino acid corresponding to the original codon or a neutral substitutes. Some eukaryote cells normally synthesis opal suppressor tRNAs. The function of these is not clear and they usually do not prevent normal termination of protein synthesis at an opal codon.

Open Circle One of the three conformations which a plasmid molecule may adopt. In an open circle there has been either a single break or a number of staggered breaks made in the strands of the duplex. Such a molecule may not form super coils and therefore is referred to a n open or relaxed circle.

Open continuous Culture A cell suspension culture with a continuous influx of fresh medium, maintained at constant volume by the efflux of cells and spent medium.

Open Reading Frame (ORF) Region of a DNA sequence that has the potential to be read as codons specifying amino acid (i.e. it is not interrupted by monense codens as top signals). ORF sequences are frequently encountered in the sequencing of gnee banks and only occasionally can they be assigned to a particular gene.

Operator The region of an operon that acts as the binding site for the repressor.

Operon Concept It is hypothesis which explains the mode of action of a gene in both active and quiescent stage.

Operon Concept It is hypothesis which explains the mode of action of a gene in both active and quiescent stage. It envisages that there are two kinds of genes, structural and controlling genes. The latter, according to this concept, are of two types, regulators and operators. The operator and structural gene together constitute an 'operon'. In the absence of the substrate (inducer), the regulator gene synthesizes a protein repressor which binds itself to the 'operator gene' thus preventing the functioning of the structural gene. In the presence of the inducer the regulator gene forms a complex with a small part of the inducer, thereby leaving structural genes to operate and transcribe mRNA.

Operon The term which Jacob, Perrin, Sanchez and Monod (1960) introduced for a group of closely-linked genes which appear to affect different steps in a single biosynthetic pathway and which appear to function as a integrated unit. With the discovery that the genes of an operon are transcribed in one messenger RNA molecule, an operon could be defined as an aggregate of genes with a single messenger.

Opine Synthetase (opine syntase) Enzyme system responsible for the synthesis of opines of which are a number of different types. The genetic information coding for opine synthetase resides in the T-DNA region of the Ti plasmid. These plasmids confer virulence to strains of Agro-bacterium, which are responsible for the crown gall disease. The T-DNA is transferred to the plant cells, where it is stably integrated into the nuclear genome and opine synthetase gene is transcribed and translated.

Opportunistic Infections A disease caused by an organism that is not usually associated with the particular host (e.g. infection of humans with dry rot fungi, of the wide range of organisms that exploit the loss of immune resistance in the disease AIDS).

Opportunity Cost The sum of the benefits an animal forfeits by not being able to perform some other behaviour during the time when it is performing a given behaviour.

Opsonic Index An *in vitro* assessment of the relative concentration of opsonins in a sample of blood. The opsonic index is the ratio of phagocytic activity in the patient's blood to that in the blood of a normal (non-immune) subject. it is determined by initially incubating both (citrated) blood samples for 15 minutes with a standard number of the relevant antigen particles 9usually microbial cells. Blood smears are then prepared form each sample and the average number of particles engulfed by the phagocytes in each sample is determined; the two averages are expressed as a ratio (the Opsonic Index).

Opsonin A substance (usually an antibody or complement fragment C3b) that adheres to bacteria. Virues, and antigen-antibody aggregated molecules and makes those particle more susceptible to ingestion by phagocytes.

Opsonization A process by which phagocytosis is facilitated by the deposition of ospinins (e.g. antibody and C3b) on the antigen.

Optic Chiasma The site in the floor of the vertebrate forebrain at which fibers entering vi the optic nerves cross to the opposite side of the brain, forming an X-shaped pattern. Inman and other mammals with good stereoscopic vision, only half the fibers from each side cross over. In lower vertebrates, nearly all the fibers may crossover.

Optical Density Degree of absorbance of light by a solution as a function of the concentration of the solute. For solution that absorb light, the relative optical density provides information about concentration an the character of the solution. Nucleic acids absorb light maximally at a wavelength of 260 nm, whereas proteins absorb maximally at 280 nm. Thus the protein content of an impure solution of DNA can be measured. Since the optical density of a DNA solution increases when the DNA is melted (the hyperchronic effect) changes in optical density can be used to provide detailed information about DNA complexity, especially in the use of reduction of optical density after melting, as is used to construct cot plots.

Optical Isomers Molecular isomers that induce or rotation of plane-polarized light when it is passed through a solution of the molecule. The rotation, which results from the asymmetric form of the molecule, may be to the right (DEXTRO) or to the left (LAEVO) for different isomers of the same molecule. These isomers are referred to as in the case of for example, glucose, as d-glucose or I-glucose.

Optimum Temperature Range A narrow range of temperature, which may cover only two or three degrees, in which cultures of an organism show maximum growth rate. Below the optimum growth rate, the rate in general declines only-slowly with the decrease in the temperature, whereas at temperatures only a few degree above the optimum a precipitous decline in rate to zero may occur.

Orbivirus A genus of Arboviruses recognized by some authors. The Orbivirus virion is aa non-enveloped icosahedral virus; 50-60 nm in diameter, which contains a fragmented linear double-stranded RNA genome-i.e. a genome consisting of 10 or more pieces; an RNA polymerase forms of the virion. Replication occurs within the cytoplasm of the host cell. Orbiviruses include the casual agents of e.g. African horse sickness and Blutongue.

Orcein Dye used in cytology, especially when dissolved in acitic acid as aceto-orcein for the staining of squash preparations of chromosomes.

Order In taxconomy, a major subdivision of a class. Each order consists of one or more related families.

Ordered Tetrad Linear sequence of the products of meotic division as visualized in the row of eight ascopores of a fungal ascus. Crossing-over events can be studied by growing each spore separately and scoring for its expression of one allele at a particular locus.

Organ Culture The maintenance or growth of organ primordial in the whole or parts of an organ in vitro in a way that may allow differentiation and preservation of the architecture and/or function.

Organ A differentiated, organized plant structure (leaf, root, flower etc.) composed of more than one tissue that performs as a functional and structural unit.

Organelle Membrane-bound structure found in eucaryotic cell containing enzymes for specialized function. Some organelles, including mitochondria an chloroplast, have DNA and can replicate autonomously.

Organelle Membrane-bound structure found in eucaryotic cell containing enzymes for specialized function. Some organelles, including mitochondria and chloroplast, have DNA and can replicate autonomously.

Organelles Organised structures that are found in or on cells. Examples, ribsosomes, nuclei, mitochondria, chloroplasts, cilia, and contractile vacuoles.

Organic Consolvent Compound used to dissolve some neutral organic substance, as in media preparation. These include alcohol (usually ethanol), acetone and dimethylsulfoxide (DMSO).

Organizer, Embrayonic A region of an embryo which directs the development of nearby regions. In amphibian early gastrulas, the dorsal lip of the blastopore.

Organogenesis The evolution, from dissociated cells, of a structure which shows natural organ form or function or both.

Origin of Life Geological evidence strongly suggests that life originated on earth about 4000 million years ago. The basic components of organic matter-water, methane, ammonia, and related compounds-were abundant in the atmosphere. Energy from the sun (cosmic rays) and lightning storms caused these to recombine into increasingly complex organic molecules. Particular combinations of such complex substances eventually showed the characteristics of living organisms. The experiments of Redi (17th century) and Pasteur (19th century) indicate that life cannot develop from inorganic matter under present day conditions.

Origin of Species The production of new types of organisms in descent from extant types. Natural selection the process by which changes are favoured leading to the production of new types. The term was introduced by Charles Darwin.

Ornithine Basic amino acid that is an important intermediate in the urea cycle.

Orphan Viruses Range of viruses that maybe found in the respiratory and digestive tracts of healthy individuals; such viruses are not associated with any known pathology. Double-stranded RNA reoviruses are examples of orphan viruses.

Orthogenesis Evolution in a definite direction and along a predetermined route, irrespective of natural selection. Since this theory implies that evolution proceeds independently of natural laws, it is impossible to test experimentally. Orthogenesis therefore conflicts with conventional theory such as neo-Darwinism.

Orthorade Transport Axonal transport form the cell body of the neuron towards the synaptic terminal. Opposite of retrograde transport and probably dependent on a different mechanochemical protein (almost definitely kinesin) interacting with microtubules.

Osmoregulation Regulation of the chemical composition of the body fluids of anorganism.

Osmosis The tendency of a fluid or solvent to pass through a semipermeable membrane from a solution of low solute concentration to one of higher solute concentration, thus equalizing the conditions on either side of the membrane. In living cells, the semipermeable membranes involved are usually the plasma membranes and/or the tonoplast and the solvent is water.

Osmotic Pressure The hydrostatic pressure generated byosmotic flow of a solvent (like water) through a semipemeable membrane into an aqueous phase containing a higher concentration of the solute. Plant cells function as osmometers, which affects water movement between plant cells and in and out of roots. Osmotic pressure is measured with an osmometer.

Osmotic Shock Passage of solvent into a membrane-bound structure due to osmosis, causing rupture of the membrane. A method of lysing cells or organelles.

Osteoclat Any of the cells that attack and erode the calcified cartilage of membrane formed in the early stages of ossification of bone. Blood vessels, preceded by osteoclasts, invade the tissue, and then osteoclsts lay down the permanent structure of bone.

Ostiole Pore in fruit bodies (pycnidia and perithecia) of certain fungi and in conceptacles of brown algae through which, respectively, spores or gametes are discharged.

Ouchterlony Technique Precipitation test involving antibody-antigen interaction in gel diffusion systems. Two or more wells are cut in an agar slab, ad a solution of antibody is placed in one central well. Other wells have various presumptive antigens in solution placed in them. By allow in simple diffusion, lines of precipitation are formed between the wells, each one representing a specific interaction.

Oudin Technique test involving the formation of zones of precipitation resulting from antigen-antibody interaction: similar to the ouchterlony technique. A gel column is prepared with an even distribution of a selected antibody molecule. Following the deposition of an aqueous suspension of antigen molecules on the upper layer of the column, lines of precipitate form as the antigen molecules move into the column and interact specifically with the antibody.

Outside-out Patch A variant patch clamping technique, in which a disc of plasma membrane converts the tip of the electrode, with the outer face of the plasma membrane facing outward, to the bath.

Ovalbumin One of the major proteins of egg white. It is the major nutritional protein of a avian egg white an is a phosphoprotein of 386 amino acids.

Overdominance Hybrids showing greater survival rate than both dominant and recessive homozygotes. Overdominance is not to confused wit heterosis in which the hybrid shows greater vigour, greater dimension etc. vigour does not necessarily mean higher survival rate because some vigorous F_1s, much more vigorous than either of the parents, are partially to completely sterile.

Overlapping Code The genetic code is in general non-overlapping, in that each triplet of bases codes for a distinct amino acid, and there is no sharing of common bases between adjacent triplets. Some overlapping genes do not provide examples of uses of the same sequence with a change in reading frame, which involves triplet overlap.

Overproducer A mutant cell that produces large quantities of chemical that is not usually accumulated. This is due to a fault in metabolic regulation. Overproucers generate large amounts of secondary metabolites or produce intermediates of central metabolism. Many industrial fermentations are carried out using mutants or strains that are overproducers.

Overwinding Situation in which the DNA double helix, already wound in a right-handed spiral, is placed under additional torsional stressing the same direction resulting in positive supercoiling of the double helix upon itself.

Oviduct A duct in animals that conveys ova from the ovaries to the exterior. It may or may not connect directly with the ovary; in vertebrates, ova are shed into the coelom before entering the open ciliated funnel at this anterior end.

Oviparity The production of undeveloped eggs, which are laid or spawned by the female. Fertilization may occur before their release, as in birds and some reptiles, or after, as inmost invertebrates, fish, and amphibians. Large numbers of eggs are usually produced because of their poor chances of survival, due to lack of maternal protection. Each egg contains a large yolk store to nourish the developing embryo. Compare Ovoviviparity.

Oxygen Reaction A reaction carried out in a reactor in which oxygen, rather than air, is introduced. In fermentation, such a reaction may use commercial-grade oxygen to provide the dissolved oxygen required for biochemical oxidation by the aerobic organism.

Oxygen Transfer Rate A measure of the rate at which oxygen passes from the atmosphere into solution. Such rates are determined using an oxygen electrode or by measuring the rate of oxidation of a solution of sulphate in the presence of a catalyst (cobalt or copper).

Oxygen Uptake Rate A measure of the actual rate of oxygen use by a culture expressed on a volumetric basis. It is dependant on the cell concentration times the average rate of oxygen consumption for that culture.

Oxygen An odorless, invisible, gaseous element occurring as O_2 and comprising 21% (volume) of the atmosphere. A component of almost all matter, it readily combines with most other elements. As a macroelement it is essential for plant and animal life. Plants produce O_2 during photosynthesis by splitting water.

Oxytocin a peptide ormone, produced by the hypothalamus and posterior pituitary gland, that acts on smooth mucscle. At the end of pregnancy the uterus becomes very sensitive to oxytocin, which promotes labor and also the release of milk from the mammary gland. It is used to induce labour artificially.

Ozone A reactive form of oxygen produced by the discharge of electricity in air or oxygen in the atmosphere by the action of solar ultra violet (UV] or by UV in laboratory sterilizing lamps. It has a faint blue colour and faint chlorine-like odour. It is used to sterilize and purity water and air and in bleaching.

P

Photosynthetically Active Radiation (PAR) The photon flux expressed as moles or microeinsteins per meter squared per second or as watt per meter squared over the wavelength range of 400-700 nm; the part of the light spectrum which is primarily absorbed by plants and used in photosynthesis.

P The probability with which a certain event may occur if the experiment is repeated under identical conditions. P=95 means that if the experiment is repeated, 95 out of 100 times the chance are that the same results would be obtained.

P_1 The first parental generation, i.e. The parents of an F_1:P_1 the grand-parents.

Pacemaker That part of the heart which undergoes most rapid spontaneous contraction, thus setting the pace for the beat of the entire heart. In mammals, the sino-atrial node. Also, an artificial device, implanted in the heart, that initiates rhythmic, contraction of the organ.

Packed Bed Reactor A tubular bioreactor that is filled with a solid matrix. It contains entrapped cells or enzymes, or is coated with cell or enzymes. It is run as flug flow reactor, with the medium through the matrix, during which time the substrate is converted to product, which emerges in the effluent from the column.

Packing Ratio Ratio expressing the degree of compaction of the DNA double helix into the fibres of chromatin that contain it. Each chromosome of the human complement of 23 in the haploid genome contains, on average, 5 cm of uncoiled DNA, this is reduced by packing to about 1 mm of 30-nm diameter chromatin fibre, and this fibre is further condensed by the higher order structure of chromatin into a chromosome of average length of only 6 ìm.

Pairing Type A group of individuals, (for instance in Paramacium) between which pairing does not occur; pairing only taking place between individuals belonging to different pairing types.

Paleobiology The attempted reconstruction-from fossil evidence, from knowledge of the properties of chemical substances, and from theories about primitive earth conditions of the early stages in the evolution of life.

Palindrome A segment of duplex DNA in which the base sequence of the two strands exhibit two-fold rotational symmetry about a centre.

Palindromic Sequence A bilaterally symmetrical DNA sequence which, therefore, reads the same in both directions. The target sites of most restriction enzymes are palindromic sequences, e.g.

Palisade Parenchyma In leaves, one or several layers of tightly packed, columnar photosynthetic cells, frequently found just below the upper epidermis.

Palmitic Acid A saturated fatty acid occurring widely in fats and oils of animal and vegetable origin.

Pancreozymin A peptide hormone released from the mucosal cells of the duodenum that stimulates contraction of the gallbladder and secretion of digestive enzymes by the pancrease.

Pandemic Affecting the majority of the population of a large region, or epidemic concurrently in many different parts of the world.

Pangenesis The theory that the heredity of organisms is determined by the summation of influences from an indefinite number of particles (pangenes) derived from all parts of the body tissue and variably affected by variation sin the environment (Darwin, 1868).

Panicle A compound raceme formed by brancing of the peduncle, each branch bearing a raceme, e.g. oat. The term is often applied to any sort of branched racemose inflorescence, for example the horse chestnut, in which each branch is actually a cyme.

Panmixis Unrestricted Random Mating. Properly excluded by any restriction, but sometimes used where random mating is assumed within the operation of a restriction, especially with dioecy in animals (Welsmann, 1895).

Papain Water soluble proteolytic enzyme [protease] extracted from papaya fruit and used especially as a meat tenderizer.

Paper Chromatography A chromatographic method using absorbent paper by which minute amounts of material can be analysed. A paper strip with a drop os test material at the bottom is dipped into the carrier liquid (solvent) and removed when the solvent front almost reaches the top of the strip. Two dimensional chromatorams can be produced using square paper and two different solvents. The paper is removed from the first carrier liquid, turned at right angles and dipped into the second. This gives a two dimensional 'map' of the constituents of the test drop. The identify of the constituents may be found by measuring the Rf values.

Paper Raft Technique A technique developed by W.H. Muir (1953) to promote development of single cells taken from suspension cultures in which cells are placed into filter paper squares set on actively growing callus (nurse tissue). Growth factors and nutrients from the callus tissue diffuse through the filter paper, promoting cell growth and development.

Para-chlorophenoxyacetic Acid A synthetic growth regulator of the auxin type which is sometimes used in plant tissue culture media. It dissolves in base.

Paracoagulation The clumping which occurs on the addition of plasma, or soluble complexes of fibrin monomers, to a suspension of bacterial cells which have been coated with protamine sulphate; the protamine sulphate thus simulates bound coagulase. A fibrin monomer is a fibrinogen molecule which, by the action of thrombin or a simplar been excised.

Paralogus Genes Homologous genes that have arisens through a gene duplication and that have evolved in parallel within the same organism (cf. Orthologous Gene).

Paramyzovirus A genous of viruses whose hosts include man, avians and other vertebrates; paramyxoviruses include the causal agents of e.g. fowl pest and numps. The Pramyzovirus virion (commonly 120 -350 nm in diameter) is an enveloped (ether-sensitive) singe-stranded RNA virus which contains a non-fragmented helicap nucleocapsid and associated RNA-dependent RNA polymerase.

Parapatric Seciatin Development of reproductive isolation among members of a continuous population in the absenc of a geographical barrier. (contrast with geographic, smpatric speciation).

Parasexual Systems Genetic systems which yield recombinants by means other than the regular alternation of fertilization and meiosis.

Parasitim A qualitative term referring to kind of symbiosis in which one party (the parasite) lives at the expense of the other (the host), contributing nothing to the relationship, and frequently destroying the host in the process; cf. Parasitization.

Prasitization A quantitative term referring to the proportion of a host population attacked by parasites; cf. Parasitism.

Parasynapsis The association of chromosomes side by side observed at zygotene and pachytene, as opposed to their alleged 91905-1935) end-to-end association at this stage, or telosynapsis.

Pratope The part of an antibody molecule which makes contact with the antigenic determinant.

Parenteral Injunction by a route other than oral, that is, intravenous, intramuscular, subcutaneous.

Parthenocapy The development of fruit without seed as the result of (a) lack of pollination, with or without artificial stimulation, (b) lack of fertilization

Partial Dominance A case where dominance is not complete and the hybrid is intermediate in type between the two parents. In F_2 1:2:1 ratio is expected in a monohybrid (cf. Incomplete Dominance).

Partition Coefficient A constant that expresses the ratio in which a given solute will be partitioned or distributed between two given immiscible liquids at equilibrium.

Passage Cells Endodermal cells that have Casparian strips but remain otherwise unthickened after deposits of lignin and cellulose have been laid lown protozylem and allow transport of alter and solutes from the cortex to the stele. They are particularly common in the endodermits of older monocotyledon roots.

Pasteurization Healting oto destroy pathogenic microbes. Methods: heating to 62.9°C for 30 minutes (holding method); or heating 71.6 °C for note less than 15 seconds. The heating is followed by rapid cooling.

Patch Test A tuberculin skin test in which the tuberculin preparation is applied to a clean skin area on a piece of gauze on tape. A positive reaction. Indicating sensitivity to the tubercle bacillus, will show an inflamed area on the skin.

Peat Partially decomposed plant material that accumulates in temperate humid climates, often forming a layer several metres deep. Peat varies from light spongy material (sphagnum moss) to a dense brown humidified material in the lower layers. If mineral salts are present in the waterlogged vegetation, neutral or alkaline fen peat is formed (the salts neutralize the acid produced by decomposition). If there are no mineral salts in the water (as in rain), acid bog peat is formed. Peat is used as a fuel and is the first step in coal formation.

PEG, Polyethylene Glycol, Carbowax This polymer is available in a range of molecular weights from ca. 1000 to ca. 6000. PEG 4000 and PEG 6000 are commonly used to promote cell or protoplast fusion and to facilitate DNA uptake in the transformation of organisms such as yeast. PEG is also used to concentrate solutions by withdrawing water from them. The polymer as the general formula.

Penicillin Resistance The property possessed by some bacteria actinomycetes, yeasts and cyanobacteria that enables them to grow in the presence of penicillin. This is often associated with the presence of the enzyme penicillin. This is often associated with the presence of the enzyme penicillinase, which is carried no some natural, wild-type plasmids.

Peptide A compound consisting of two or more amino acids covalently linked. Peptides with three or more amino acids are pllypetides.

Peptidyl Transferase Enzyme that catalyses the movement of amino acid-charged tRNA molecules (aminoacyl-tRNA) on the ribosome during the translation stage of protein synthesis. The charged tRNA molecules engage consequtively with the A site and the P site on the ribosome prior to discharge, and the transferase facilitates the movements of the tRNA from one site to another.

Percolating Filter A system of aerobic wastewater treatment based on a plug flow fixed film bioreactor.

Pericentric Inversion A chromosomal inversion of 180° in which the inverted segment includes (surrounds) the centromere.

Perinatal Pertaining to the period of childbirth and shortly thereafter; in medical statistics, the period usually begins with the birth of a fetus 20 weeks or more of gestation and ends 7 to 28 days later.

Pest Upset A condition in which innocuous species become manmade pests as a result of insecticide use which unintentionally destroys their population-regulative natural enemies.

Pest A disease-inducing organism, often applied to an organism that causes disease or attacks agricultural crops.

Petite A slow-growing strain of yeast (Saccharomyces) lacking certain respiratory enzymes and forming unusuallysmall colonieson agar, (sometimes called vegetative petities) bear mutant mitochondrial DNA.

PFA (Plaque Forming Cell) An antibody producing cell detected *in vitro* by its ability to lyse antigen-sensitized erythrocytes in the presence of complement.

pH Gradient Gradient employed in techniques such as iisoelectric focusing, in which proteins migrate to positions equivalent to their own insoelectric point.

Phage Induction Transition from the prophage state lytic growth of a temperate phage in response to an external stimulus. Induction results in the initiation of transcription of phage genes, excision of the prophage from the host chromosome, and the synthesis of phage genomes and capsid proteins.

Phagocytosis A form of endocytosis; the uptake of a solid particle by forming a pocket of plasma membrane around the particle and pinching off the pocket tc form an intracellular particle bounded by membrane.

Phase-contrast Microscope An instrument that translates differences in the phase of transmitted or reflected light into gradations of contrast with background.

Phenocopy A non-hereditary phenotypic change that is environmentally induced, during a limited developmental phase of an organism, that mimics the effect of a known genetic mutation.

Phenol Coefficient A test for evaluating the effectiveness of a disinfectant by compaing it with phenol under identical conditions.

Phenotypic Flexibility The ability of a genetype to vary its expression in different environments, of twice or three times as great as those causes by the genotype, is a character of considerable adaptive significance in most plants. Species in a genus as also populations within a species show differences in the phenotypic flexibility, indicating that this attribute is genetic based and subjected to the control of the usually processes of natural selection.

Phoresy A kind of symbiosis involving the use of one species by another for transport.

Phosphagen In excitable tissue, an energy-string compound containing a high-energy phosphate group, usually in enzymatic equilibrium with the terminal phsophate of ATP.

Phosphate Triester Synthesis A technique used in the production of oligonucleotides which uses protected monotriethylammonium salts of nucleotides. In the presence of a coupling agent, such as mesitylenesulphonyl nitrotriazole, the triesteris incorporated into the growing ching of an oligonucleotide.

Phosphatidic Acid A simple glyceride composed of two long chain fatty acids esterfication to a molecule of glycerol phosphate.

Phosphogluconate Pathway An oxidative pathway beginning with glucose 6-phsphate and leading via 6-phosphogluconate to NADPH, pentoses, and other products.

Phosphoglyceride A fatty acid ester of glycerol phosphate with an alcohol group esterfied to the phosphate.

Photoauxotroph Organism that requires only light plus inorganic compounds to provide all of its energy and nutritional requirements.

Photosynthesis Metabolic processes, carried out by green plants, by which visible light is trapped and the energy used to synthesize compounds such as ATP and glucose.

Photosynthetic Efficiency A measure of the extent to which light energy intercepted by a leaf, canopy or photosynthetic cell suspension is converted into organic matter. Efficiency is expressed in term of total incident light or photosynthetic active radiation (PAR).

Photosynthetic Productivity It is an index of the yield, in terms of harvested dry weight, produced by a given crop or within a given ecosystem. Productivity is usually expressed in terms of tonnes dry weight produced per hectare per year.

Phylogenetic Series An evolutionary chain of organisms, most of the members of which are usually extinct, which indicates the descent of present-day types from earlier forms.

Physical Containment The use of physical barriers to prevent the escape of genetically engineered organisms from the laboratory into the environment. Precautions employed include the maintenance of negative air pressure within the laboratory and they use of fume hoods with HEPA filters on their exhausts.

Phytoalexins Substances produced by a wide range of plants in response to bacterial or fungi infections or, less commonly, following chemical exposure or injury. The are non-specific antibiotic.

Phytohaemagglutinin It is a lectin which binds to surface glycoproteins of some animal cells stimulating cell division.

Phytotoxin Bacterial toxins .hat affect plants.

Phytotron A controlled environment Building.

Pigment A coloured compound; as chlorophyll in green plants. The amount and type of coloured material is referred to as pigmentation.

Pilot Plant An intermediate production facility used to study scale-up and other production problems between bench-scale experiment and full commercial production.

Pinocytosis A form of endocytosis; the uptake of liquids by enguilfing sample of the external medium into a pocket of the plasma membrane followed by pinching off the pocket to form an intercellular vesicle.

Pistill The entire part of the flower that produces megaspores (which, in turn, produce eggs). Sometimes, and incorrectly, referred to as a female floral part.

Planetary Quarantine An area of endeavour involved in the prevention of contamination of other planetary bodies and the moon by terrestrial organisms, as well as prevention of contamination of earth by organism from extraterrestrial bodies.

Plasnosome A supernumberary chromosomes resulting from non-disjunction of a bivalent in meiosis.

Plant Cloning Vehicle It is vector used in gene manipulation to carry foreign DNA and insert it in an inheritable manner into plant cells, Possible vectors in include the Ti plasmid of Agrobocterium tumefaciens and the DNA plant virus.

Plant Nutrient An essential plant element required in large (macronutrient) or trace (Micronutrient) amounts.

Plant Row A row of plants that represents the random progeny from a single plant following a cross. For example, F_3 plant rows would be all the progeny from a single F_2 plant.

Plaque (viral) A clear area in a single layer or monolayer of cells created by viral lysis of infected cells.

Plaque Hybridisaton An adaptation of the technique of Southern hybridisation. Bacteriophage plaques of an agar plate are transferred to a nitrocellulose filter, or other suitable DNA-binding matrix, by pressing the filter gently onto the surface of the agar plate. A small amount of the phage and unpackaged phage DNA binds to the filter leaving behind the original plaques. The phage DNA is denatured on the filter which is then baked to attach the DNA firmly. The bound DNA can now be hybridized with a probe, usually ^{32}P-labelled DNA or RNA. Any homology between the probe and the DNA on the filter will allow the two species to anneal and the hybrids formed can be detected by autoradiography. This technique is applicable where a chimeric phage particles is carrying the gene of interest. In such a situation, a bacterial lawn is infected with a mixture of chimeric phageparticles (i.e the library) and a large number of plaques develop overnight. These plaques can be treated just lie the colonies in colony hybridization to identify and isolate the chimeric phage particle carrying the gene of interest.

Plasmagene A particle whose reproduction in the cell, but outside the nucleus, and transmission by the egg (and sometimes by individuals to their parents. Cf. Virus. May be taken to include Plastogene.

Plasmid Vector A plasmid involved in the transfer in a gene or genes carried on a length of foreign DNA into a host in which they do not normally occur. This is one of the techniques used in what is known as in vitro genetic manipulation, gene (genetic) engineering or genetic manipulation. The plasmid used is a vector or a cloning vehicle.

Plasmids A plasmid is a replicon that is stably inherited (maintained without specific selection) in an extra chromosomal state. Most of the plasmids are not required for the survival of in which they reside. In many cases, however, they are essential under certain environment, such as in the presence of an antibiotic such as chloromphenical, ampicillin, tetracycline, kanamycin etc. In Rhizobium leguminosarum, the genes for nitrogen fixation and nodule formations are located in plasmid.

Plasmolysis Contraction or shrinking of a cell's cytoplasm due to the loss of water by osmotic action, i.e., a loss of water molecules.

Plating A technique used to obtain pure cultures of microorganisms (bacteria, yeasts and fungi) that produce a distinct colony when grown on a solid medium (a nutrient medium solidified with agar or similar agent in a petridish).

Plug Flow Reactor This reactor is ofentubular in structure, in which the substrate passes through an essentially unmixed system which may consist of anunstairred liquidphase of a system in which the biocatatlysts consists of a stationary matrix such as immobilized cells or enzymes, or packed cells.

Pluteaceae Basidiocarp: stipitate with a volva, the lamellae each having a convergent Bilateral Trama; basidiospores:pink, Genera include Volvariella.

Pock formation this phenomenon occurs in Streptomyces. A zone of inhibition of psorulation and retardant of growth of a plasmid-free strainis formed around the colony of plasmid-containing strain. The term is also known as Lethal Zygosis.

Point Mutation A nutation that results from a small, localized alteration in the chemical structure of a gene. Such mutation can give rise to wild-type revertants as a result of reverse mutation. In genetic crosses, a point mutation behaves as if it resided at a single point on the genetic map. (contrast with deletion.).

Poky Mutant A mutant strain of Neurospora in which the rte of growth is lower than in wild type strins; the reduced growth rate is associated with the absence for deficiency of certain components of the respiratory chain.

Polar Fusion Nucleus Equivalent to diploid fusion nucleus.

Polar Granule Granule found in the posterior part of the developing insect egg. That is rich in RNA. In drosophila only cells that have formed following the engulfment of some of these granules have the potential to form germ cells.

Polar Mutation A mutation of one gene that affects the expression of the adjacent nonmutant gene on one side, but not of that one the other side.

Polarization Of chromosomes at telophase of mitosis and, later, the maintenance of their prodimal parts on the polar side of the nucleus.

Poliovirus smell enerovirus, with a single strand of RNA as genome and an icosahedral capsid. It is the causative agent in poliomyelitis.

Pollen Culture Cultivation of pollen grains in vitro for producing haploid plants (generally referred to an anther Culture).

Polyaenylation The nontrancriptive addition of poly a to the 3'-end of eucryotic RNA.

Polycistronic mRNA A single mRNA molecule produced by the transcription of more than one cistron.

Polycross Progeny Progeny from a selection, line or clone outcrossed to other selections rowing in the same polycross nursery.

Polyethylene Glycol (PEG) or Carbowax A fusion-inducing agent (fusogen) for agglutinating protoplasts which is used in somatic hybridization studies. This compound is also sometimes used in media as a non-metabolite osmoticum and is available in various molecular weights, ranging from ca. 200 to 6,000.

Polygedron The crystal-like inclusion body, characteristic of infection by a polyhedrosis virus, the variously-shaped protein aceous coverignof the virions produced in cells infected by nuclear or cytoplasmic polyhedroses.

Polygenic Character A character whose inheritance is controlled bymanygenes eachhavingan effect which issmallcompared with non-heritable variation.

Polygenic Trait (Quantitative Inheritance) A trait not readily separable into discrete groups but which may be continuously variable and often fits a normal distribution curve. Such a trait is controlled. By more than one pair of genes and may also be modified by environmental factors.

Polymorphic Two or more recognizably different sorts of individuals within a single breeding population.

Polypeptide Chain A chain of amino acids whose exact synthesis is under the control of DNA. Some proteins consists of only one chain; others are aggregates of several chains.

Polyploiding Agent A substance or treatment which, under given circumstances, is capable of inducing poyploidy. The following substances have been used successfully; Acenaphthene and various derivative thereof; Apil; Aurantia; Colchicine: diphenylamine; Easters of naphthoic acid: Ethyl mercuric chloride in conjunction with 1:2:5;6-dibenznthracene; Naphthaleneacetic acid: Paradichlorbenzence: Sanguinrine; Sulphanilamide; Veratrine suplhte.

Polysomic An otherwise diploid individual having on chromosome represented three (Trisomic) or four (Tetrasomic or doubly Trisomic) times instead of twice (2x + 1, 2x + 2).

Polytene Nucleus One in which protein production and gene reproduction go on side by side. A permanent stage in which the chromosomes are represented by many linearly stretched and paired threads in the Diptera, especially in the salivary glands.

Population Cross A cross between two open-pollinated varieties (Syn., Varietal Cross).

Population Doubling Level The total number of population doublings of a cell line or strain since its initiation in viro a formula to use for the calculation of "population doublings" is the following: number of population doublings = Log 10 (N/NO) x 3.33.

Population Genetics The story of genetic principles of heredity at the population level.

Position Effect (cis-trans, Lewis Effect) The phenomenon whereby different phenotypes result from the combination of pseudoalleles depending on whether they are in the cis (same homolog) or trans (different homologs) position: modification of the effect of gene by its translocation to another chromosome: Postmating Isolating Mechanism Any intrinsic isolating mechanism (for example hybrid sterility) that operates after mating.

Potency (of a gene) The capacity of a gene for manifesting its presence cf. Expressivity and Penetrance. Partial potency is synonomous with incomplete dominance.

Power Number A dimensionless group of relevance to the stirring of fluids in termenters that represents the intertial forces transmitted to the fluid.

Prader Willis Syndrome Chromosomally based congenital disorder of humans resulting from a small deletion in the long arm of chromosome 15 (in over 50 percent of affected individuals only). Suffers are mentally retarded, with small hands and feet, some narrowing of the head and a tendency for obesity to develop.

Pre-adaptation, Constitutional Adaptation in which a species is, by its existing peculiarities, already modified to suit certain modes of life or particular types of environment.

Precipitin Test or Microprecipitin Test A serological assay in which visible particulates form when soluble antigen and antibody react. This precipitin reaction detects and identifies antigens.

Preformation The concept that the individual is present in miniature in the embryo and that development to adulthood involves growth but not differentiation.

Presumptive Hybrid A cell or individual plant for which there is evidence to suggest that is a hybrid as somatic cell fusion products intermediate in colour to that of (stained) parental cell lines.

Prezygotic Incompatibility In the case of incompatible or wide crosses the inhibition of pollen germination or the prevention of pollen tube growth, among other possible barriers to plant fertilization. In some cases these barriers be can be overcome through in vitro pollination and/or embryo culture.

Primary Growth (1) Refers to apical meristem derived growth; the tissues of a young plant. (2) Refers to growth during the initial culture period as primary callus growth.

Primary Response The rather weak response of the immunological system upon its first exposure to a given antigen, characterised by first lgM and then lgG synthesis.

Primer DNA An asymmetric piece of DNA which acts as a template and a primer for the in vitro synthesis of DNA by DNA polymerase.

Primer RNA Short RNA sequence of about 10 nucleotides that is synthesized by a primase on the lagging strand during DNA replication. These short lengths of RNA are then extended by DNA polymerase to yield the oazaki fragments, from which the primer

RNA molecules are then erased and the pieces 'stitched' together by DNA ligase to yield a continuous new DNA strand.

Primosome Protein complex that is involved in the primary action to produce the okazaki fragments of DNA replication. The complex includes a primase that synthesizes the primer RNA.

Probability Function The function relating observations to the parameter from whose value their frequencies maybe predicted. Cf. Likelihood function.

Probability The likelihood of occurrence of a given event. Usually expressed as number between 0 (complete certainty that event will not occur) and 1 (complete certainty that the event will occur).

Probe DNA DNA made highly radioactive for use in molecular hyubridization studies. The radioactivity is usually introduced either during synthesis by reverse transcriptase or, in the case of an existing fragment, by nick translation. Procaryotic Cell A type of cell in which the nuclear substance is not enclosed within a membrane, e.g., bacteria and blue green algae.

Processive Enzyme Enzyme that continues to act on a substrate even during repetitious catalytic event and release of product.

Prodormal Period A state in infectious disease process in which the first symptom appear.

Product Recovery Membrane Fermentation An experimental system that has been proposed for continuous ethanol recovery from fermenter broth employing selective membrane separation techniques.

Promeristem or Protomeristem The embryonic meristem-containing organ initials or foundation cells.

Promiscuous A term used to describe a broad host range trans missible plasmid.

Promoter In the DNA of an operon, a nucletode sequence that is recognized by RNA transcriptase as a site at which to begin transcription.

Prosthetic Group A heat stable metal ion or an organic group (other than an anino acid) that is bound to a protein and serves as its active group.

Protein Sequencing It is an analytical method used to determine the sequence of amino acids which make up peptide or protein.

Protein One of a class of complex organic nitrogenos compounds composed or an extremely large number of amino acids joined through peoptide linkages.

Protelean Parasite An insect species in which only the immature stages are parasitic.

Protoplast A cell which would normally have a cell wall, but from which the wall has been removed by enzymatic digestion or by special growth conditions.

Protostomia Those animal group in which the blastopore becomes the mouth.

Protozoa A group of single-celled organisms classified by some biologists as a single phylum; includes the flagellates, amoebs, and ciliate. In modern classification protozoans are elevated as distinct kingdom (Protista) and each major subgroups to the rank of phylum.

Protroph Micro organisms which can grow in a minimal medium consisting of a sugar, mineral nutrients and vitamin e.g. biotin. They can synthesis the medium. Recently this term is also being used to describe biochemical mutants in Arabidopsis thaliana, a Cruciferous plant.

Pseudo Alleles Genes which behave as allelic in a small population but show a very small percentage of recombination in a very large population. Pseudo alleles, therefore, refer to genes or sites in a locus which are very closely linked but separable through recombination.

Psudo-plastic Fluid A fluid in which the apparent visocosity decease, but subsequently becomes linear. Such behaviour is characteristic of many cultures of moulds or actinomycetes.

Pseudomonad Any genus of polarly flagellated, gram negative, rod shaped bacteria, thee include the fluorescent pseudomonads, which produce a water soluble, greenish yellow fluorescent pigment and the marine luminous bacteria.

Pseudopolyploidy Apparent polypooidy that has arisen by single or repeated transverse division of the chromosomes in species with diffuse centromeres.

Pst I (Pronounced Pisst One) A type II restriction enzyme from the bactriu Providencia stuartil 164 which recognizes the DNA sequence.

Pulse Chase Experiment A radio actively labelled compound is added to living cells or extract (pulse) and, a short time later (e.g., five seconds), an excess of unlabelled compound is added to dilute out the "hot" compound. Samples are then taken at periods after the pulse to follow the course of the label as the compound is metabolized (chase).

Pure Breeding Line A strain of individuals that are homozygous for all genes being considered.

Pure Breeding A breeding method employed in domestic animals that has as an objective the preservation of the characteristic and valuable properties of certain race or breed.

Pulse Labelling Growth or reaction in the presence of a labeled compound for a short period and then examining the fate of the label.

Pure Culture Culture containing only one species, or type of microoranism.

Putrefaction A decoposition, catalyzed by the growth of microorganisms. It is characterized by the formation of evil smelling end products which are generated as a result of the breakdown of proteins.

Putrescrible Wastes Wastes which consist mainly of plant or animal residues and which undergo degradation by bacteriological action e.g. abattoir refuse, animal residues, cannery wastes, fats, fish residues, fruit wastes, vegetable wastes, waxes – animal and plant.

Pyramid of Energy Graphical representation of the total energy contents at different trophic levels in an ecosystem.

Pyramide of Biomass Graphical representation of the total masses at different trophic levels in an ecosystem.

Pyridine Linked Dehydrogenases Dehydrogenases requiring as a coenzyme either one of the pyridine coenzymes NAD or NADP.

Pyridine Nucleotide A nucleotide coenzyme containing the pyridine derivative nicotinamide i.e. NAD or NADP.

Pyrimidine Single ring nitrogenous bases, constituents of both DNA and RNA pyrimidines in DNA are cytosine and uracil. Some unusual pyrimidine bases found in tRNA are pseudouracil, dihydroxyuracil.

Pyrophoshatase An enzyme hydrolyzing inorganic pyrophcshate to yield two molecules of (ortho) phosphate.

Q

Q Banding Chromosome banding technique so called because it involves the use of quinacrine dyes (quinacrine mustard or quinacrine dihydrochloride). The technique is most useful in demonstrating areas of heterochromatin, especially the *Y* chromosomes.

Q Enzyme Enzyme involved in the formation of (1-6) linkages at amylopectin branch points.

QH1-Cytochrome c Reductase Membrance-bound complex in the mitochondrial inner membrane, responsible for electron transfer from reduced coenzyme Qto cytochrme c. Contains cytochromes b and c_1, and iron-sulphur proteins.

Quadruple Diploid An otoploid carrying in it somatic nuclei a double set of each of fur specifically distinct genoms.

Qualitative Analysis An analysis which seeks to determine what components are present in significant amounts in a substance or mixture of substances; such analysis is designed to indicate the presence or absence of specific chemical species, and not their respective quantities.

Quantitative Analysis An analysis which seeks to determine the relative amounts of significant compounds present in a substance or mixture of substances. There is no sharp demarcation between qualitative and quantitative procedures.

Quantitative Characteristic A continuously varying characteristic; a measuring instrument is often used to distinguish quantitative characteristics.

Quantum Number A measure of the number of light quanta required for a given photochemical reaction. In photosynthesis, it is a measure of the number of light quanta required for the assimilation of one molecule of carbon dioxide or evolution of one molecule of oxygen. The theoretical minimum quantum requirement for photosynthesis. If he two photoreaction Z scheme is correct, is eight. The quantum number is determined experimentally as the reciprocal of the quantum yield.

Quantum This is the name given to a packet of radiant energy of value *hv* joules or *hv*/1. 6 × 10^{19} electron volts where *h* is Planck's constant and *v* is the frequency of the radiation. This unit as introduced by Planck in 1900. An alternative name usually used when light energy is being considered is photon. The energy in a quantum of radiation having a wavelength 5600 A is about 3.55 × 10^{19} joules so that a 100 watt bulb which emits 5% of its energy is visible light gives approximately 1.4 × 10^{19} visible quantsa per second. .

Quarantine Block An isolated block in which all plants imported into the area are first grown to observe any disease or pest that may be carried by the plants in question and prevent their further spread to those grown in the locality.

Quaternary Hybrid A hybrid derived from four grand-parental types all of which are different.

Quellung Reaction Capsular swelling test; Neufeld's test; when a specific antibody in blood serum combines with its specific capsular antigen, it gives the appearance of the capsular swelling that is typing of Strep-toccus pneumoniae.

Quelling Raction Swelling of the capsule surrounding a bacterium as a result of interaction with anticapsular antibody, consequently the capsule becomes more refractile and conspicuous.

Quenching A process that reduces the number of photons that impinge upon the window of the photomultiplier of a scintillation counter. Quenching results in a counting rate (counts per minute, cpm) that is lower than the true disintegration rate (disintegrations per minute, dpm). The difference is due to dilution effects, or effects of coloured quenchers of chemical quenchers.

Quick Stop Type of bacterial mutant in which DNA and cellular replication cease abruptly at 42 °C. Compare Slow Stop.

Quiescent Culture Culture of cells in which active cell growth and replication are minimal, but in which cells continue to survive adequately. Confluent cell cultures may behave is damaged.

Quiescent Culture Culture of cells in which active cell growth and replication are minimal, but in which cells continue to survive adequately. Confluent cell cultures may behave in this way.

Quinacrine Derivative of acridine used as a n antimalarial drug and also in cancer chemotherapy. Quinacrine dyes (quinacrine mustard or quinacrine hydrochloride) are also used as chromosome stains and will fluoresce under ultraviolet light.

Qulescent Centre A group of cells in the centre of the apical meristem in which mitotic divisions are rare or absent. The cells may begin dividing if another part of the meristerm is damaged.

Qulescent Centre A group of cells in the centre of the apical meristem in which mitotic divisions are rare or absent. The cells may begin dividing if another part of the meristem is damaged.

R

R Loop A region of a double-stranded DNA molecuie in which one DNA strand is displaced and paired with a complementary strand of RNA.

R-Factor a transferable plasmid that carries the information for resistance to antimicrobials or other agents.

R_1 Designates regenerant from in vitro culture; as a shoot from callus or a plantlet from a somatic embryoid. R_2, R_3 are the second and third generations from crosses between the R_1, R_2 generations respectively.

Rabbit Reticulocyte System A cell-free system prepared from lysed rabbit reticulocytes which is able faithfully to translate eukaryotic mRNAs from a wide variety of heterologous sources.

Rabi The winter season.

Rabies An acute disease of humans and other animals, caused by a virus Transmitted by direct contact from infected animals to humans through bites.

RAC (Pronounced Rack) Recombinant DNA Advisory Committee. A committee of the United States' NIH which formulates guidelines governing the performance of genetic engineering experiments.

Radial Symmetry The condition in which two halves of a body are mirror images of each other regardless of the angle of the cut, providing the cut is made along the centre line. Thus, a cylinder the cutis made along the centre line. Thus, a cylinder cut lengthwise down its centre displays this form a symmetry (Contrast with bilateral symmetry).

Radioactive Tracer Rèfers to definite quantity of radioisotope introduced into a biological system so that its path through the system and its concentration in particular areas can e determined by the measuring of radioactivity.

Radioimmunoassay A method for finding the concentration of peptide hormones in a biological system. Radioactive isotopes are used to label hormones which are then introduced, with specific antibodies, into the tissues or fluids that are being analyzed. The proportion of labeled hormone that becomes bound to the antibody is recorded. This figure shows a negative correlation with the amount of unlabelled hormone in the sample since the unlabelled hormone is competitively inhibiting the binding of labeled hormone. The concentration of unlabelled hormone can be found by comparing the degree of competitive inhibition it produces with a set of standards.

Radioisotope A radioactive isotope of an element. Examples are carbon –14 and tritium.

Radiology The science dealing with X-rays and other radioactive emissions and their use in biology and medicine for the examination or photographing of organs in particular.

Radionuclide An unstable atomic species characterized by a given mss number, atomic number and energy state, that decays emitting radiation (alpha, beta and/or gamma-rays).

Random Genetic Drift Variation in gene frequency from one generation to another due to chance fluctuations.

Random Mating The situation within a population in which, subject to known restrictions of determinable effect, such as Dioecy, Incompatibility and intrauterine copulation, each individual has an equal chance of mating with any individual, including itself.

Random Sample A sample obtained in such a ay that each individual in the population, or each gene in the genome, has the same chance of being selected.

Randomized Blocks An experimental design in which the treatments are arranged in random order within the blocks or replicates.

Rapid Lysis Mutant Types of *E coli* mutant in which infection by T even phage leads to particularly fast cell lysis.

Recalcitrant Molecule A compound that is slowly mineralized or that is not metabolized by micro-organisms in nature.

Receptive Field Of a neuron, the area on the retina from which the activity of that neuron can be influenced.

Receptor Sites The specific sites of the host, particularly bacterial, cell wall to which the infecting virus particles get desorbed.

Recessive Gene When the parents with contrasting characters are crossed, the character of one aren't does not appear in F_1 generation even through genes for both the factors are present in the hybrid. The gene which does not appear is said to be recessive.

Reciprocal Chiasata Chiasmata formed between two chromatids in such a way that a double crossing-over occurs in which the other two chromatids are to involved.

Recognition Site The nucleotide sequence to which a restriction enzyme binds initially.

Recombinant DNA Technology The application of genetic tools (restriction endonucleases, plasmids and transformation) to the production of specific proteins by biological "factories" such as bacteria.

Recombinant DNA A DNA molecule (in practice generally a bacterial plasmid) which has been enzymatically cut and DNA of another individual of the same or a different species inserted in the space so produced, then reannealed to the closed, circular form, Many recombinant DNA molecules, therefore, can be so constructed that an individual (e.g., a bacterium) ha a genetic capability not previously associated with that organism.

Recombination Factor The gametic output in a hybrid of new combinations of two genes expressed as a fraction or percentage of the total gametic output of new and old combinations combined. The formulae for the gametic output of the double heterozygote $AaBb$ are (i) Coupling: $1/2/1\text{-}p)AB$. $1/2pab$, $1/2paB$, $1/2(1\text{-}p)ab$; (ii) Repulsion: $1/2pAB$, $1/2(1\text{-}p)Ab$, $1/2pab$, where p represents the recombination fraction. Where there is no linkage, P 0=.5

Reducer In biology, a heterotrophic individual which utilizes the chemical energy of organic matter while breaking it down or reducing it to more simple substances.

Reductional Separation Separation of homologous parts of non-sistr chromatids at anaphase of one or other meiotic division, as proposed to equational separation of sister parts of chromatids.

Refuse Incinerator A fuel-furnace for the combustion of organic wastes, in which all gases will have reached a minimum temperature of 1350°F before discharge.

Regression Coefficient The rate of change of the dependent variate on the independent, variate.

Regressive Curve curve that expresses the change in rte of a dependent variable, with respect to an independent variable.

Regulon A system of more than one operon controlled by a single regulatory protein.

Rejuvenation Synonymous with dedifferentiation. (2) Treatment that leads to culture invogoration (such as subculture) or revival (dormancy breaking).

Relaxed Control Situation in which the numbers of a particular plasmid present in a bacterial cell are not closely regulated by coupling of plasmid DNA synthesis to host cell **DNA Synthesis** In relaxed control upto 100 copies of plasmid per cell may be found; this may be further increased using chloramphenicol.

Relative Sexuality The ability of gametes to function as male gametes in certain unions and as female gametes in other unions.

Relaxed Control Situation in which the numbers of a particular plasmid present in a bacterial cell are not closely regulated by coupling of plasmid DNA synthesis to host cell DNA synthesis. In relaxed control upto 100 copies of plasmid per cell may be found this may be further increased using chloramphenicol.

Releasing Factor One of several hypothalamic hormones that stimulate the secretion of anterior pituitary hormones.

Renner Effect Competition among the four genetically different spores formed by one meiosis in regard to which shall form the embryo sac.

Repetitious DNA DNA in which the base sequences are repeated many times in the genome of any cell: in highly repetitius DNA the base sequences may be represented more than 1000 times. Such DNA is concentrated in heterochromation and in chromosomes where it is located in the region of the centromere. Moderately repetitious DNA includes the genes for rRNA and tRNA, as well as structural genes coding for the synthesis of histones.

Repetitive DNA Repeated sequences of nucleotides that ay occur in great numbers of reiterated copies in a chromosome complement.

Replacement Substitution A change in a gene leading to a change in the amino acid sequence of hepolypeptide for which gene codes (contrast with silent substitution.).

Replacement Vector A phage cloning vector in which a portion of the phage genome is replaced by the foreign DNA fragment to be cloned. Replacement vectors are usually based on phage.

Replicator A NA segment which contains an origin of replication and is bale to promote the replication of a plasmid DNA molecule in a host cell.

Repressible Enzyme An enzyme whose synthesis can be decreased or prevented by the presence of a particular compound.

Reproductive Isolation inherent blocks to crosses between members of different breeding populations.

Reproductive Stage The stage in the life cycle of n organism in which sexual or asexual reproduction occurs.

Research Kit A package that contains all the required reagents, in a suitable form, to carry out one or more specific assays which may be based on immune, chemical or enzyme-catalyzed reactions. A research kit uses a procedure or reagents that have not been validated or accepted by health authorities.

Resistance Gene a gene which determines resistance to some inhibitor or toxin, e.g. an antibiotic or heavy meta. Resistance genes are frequently included in vector molecules since they enable the selection of organisms containing the vector.

Respirator A conventional device covering the nose and mouth, which provides a filter for inspired air.

Respiratory Pigments Coloured compounds that can combine reversibly with oxygen. Haemolobin is the blood pigment in all vertebrates and a wide range of invertebrates. Other blood pigments such as haemoerythrin (containing iron) and haemocyanin (containing copper) are found in lower animals, and in many cases are dissolved in the plasma rather than present in cells. Their affinity for oxygen is comparable with haemoglobin, thought oxygen capacity is generally lower.

Resting Phase Inter phase; the phase in the nuclear cycle when it is at rest in the sense of not dividing.

Restitution Nucleus The cell nucleus arising at meiosis owning to the chromosomes not separating into two groups at anaphase, but on account of the presence of univalents or for other reasons all being included within the same nuclear membrane. In this way the halving of the chromosome number is prevented, and the formation of restitution nuclei therefore, as a rule, leads to the origin o unreduced gametes.

Restorer Gene A gene, usually dominant, that effectively overcomes the effect of male sterile cytoplasm on male fertility, i.e., produces functional male gametes evening the presence of the male sterile cytoplasm.

Restricted Refers to the bacteriophage mutants that are capable of infecting only a few bacteria.

Restriction Endonucleases in Genetic Engineering In gene cloning experiments DNA molecules have to be cut in a very precise and reproducible fashion. Thus, endonucleases play an important role in cutting the desired gene as well as cleaving the vector. The required DNA fragment from a large DNA molecule should be cleaved in a precise manner for further genetic manipulations. A particular restriction endonuclease can recognize and bind to specific base sequence of the DNA and then will cleave it. It is highly reproducible and can be programmed according to DNA sequences of required gene and particular endonuclease identifying and cleaving it.

Restriction Enzyme An endonuclease that recognizes specific nucleotide sequences in DNA and then makes a double-strand cleavage of the DNA molecule. The DNA of an organism possessing such as Enzyme is usually modified at the recognition sites, to prevent self-cleavage, by a modification enzyme that recognizes the sites and methlates specific nucleotides at each site.

RF Value In paper chromatography, the distance traveled by the solute divided by the distance traveled by the solvent front, the latter always being taken as one. The Rf (relative front) value is thus always between nought and one and is characteristic of a particular molecule. In this way various amino acids, chlorophylles, etc., may be identified.

Rhenumatoid Arthiritis Common disease of humans affecting about one in 50 of western populations. There is a weak hereditary element and some association with specific HLA antigens.

Rhizogenesis Root formation and growth; as in root development de *novo* from callus.

Rhizome A stem that grows horizontally below ground. Rhizomes may be fleshy (e.g. iris) or wiry (e.g. couc grass) and may serve as an organ of perennation or vegetative propagation, or occasionally both. Compares tolon,

Rhizosphere The soil region subject to the influence of plant roots and characterized by a zone of increased microbiological activity.

Riboflavin or Lactoflavin [$C_{17}H_{20}N_4O_6$ mw 376.36]. A water soluble, vitamin [B_2] essential to cellular respiration. It is important in carbohydrate metabolism and implicated in pohotooxidation of auxins and perception of phototrophic stimuli. Riboflavin is an occasional addition to plant tissue culture media. It is sometimes called vitamin G.

Ribonucleoprotein (RNP) Complex of RNA and protein. Particles of RNA occur in both nuclei and cytoplasm of eukaryotic cells, and are formed in order to protect RNA from degradation and probably also to assist in transport and localization within the cell.

Ribozyme An RNA with enzyme-like activity. I is an internal guide sequence (IGS) contained within an intron of a eukaryotic transcript which is able to catalyse the splicing of that intron in vitron in the absence of any enzyme or other protein molecule.

RNA Primase enzyme that synthesizes the short primer RNA of about 10 nucleotides that is elongated by DNA polymerase to form an okazaki fragment of DNA during DNA replication. The leading strand of DNA is presumed also to be nitiated by such a primer.

RNA Sequence Analysis The determination of the base sequence in an RNA using enzymes in a method directly analogos to the Maxam-Gilbert DNA technique, except that base-specific ribonucleases are employed that cleave specifically next to G,A, or A and U, are specific for pyrimidines.

RNA Virus A virus in which the genetic information is carried in RNA. These include the picrnaviruses, the aboviruses and the myzoiruses.

RNA-Coding triplets Triplets found on mrna that code for the amino acids, the bode being recognized by the anticodon present on the molecule of tRNA charged with the specific amino acid.

RNA-dependant DNA polymerase A group of enzyme catalyzing formation of DNA molecules from RNA templates. These occur in some viruses (e.g. those that produce tumoıs).

Rodac Plate A plastic culture plate so designed that when filed with sufficient agar culture medium, a meniscus is formed which rises above the rim of the plate. The plate can then be used to assay the microbiological contamination of smooth surfaces (tabletops, floors, skin) by pressing the raised agar against the surface. A number of microorganisms are transferred to the agar. And after suitable incubation visible colonies develop. The number of colonies appearing is a reflection of the degree of contamination of the surface.

Roller Bottle Apparatus A culture system that is used for the propagation on anchorage-dependant cells or liquid cultures of aerobic organisms the culture is placed in a sealed bottle that is placed horizontally on a set of revolving rollers which, by turning the bottle slowly round on its longitudinal axis, keeps the particles in suspension.

Root Culture Isolated root tips of apical or lateral origin may produce in vitro root systems with indeterminate growth habits. These were among the first kinds of plant tissue cultures (White, 1934) and remain important research tools in the study of developmental phenomena; mycorrhizal; symbiotic and plant-plastic relationships.

Root-nodule Bacteria Bacteria belongingto the rhizobium. Family Rhizobiaceae, which live symbiotically in the nodules of roots of leguminous plants and fix atmosheric nitrogen.

Rootstock A vertical usually short, underground stem, It is found in many angiosperms, e.g. rhubarb and straw berry, and in certain pteridophytes, e.g. Osmunda. It can reproduce vegetatively.

Rotating Culture or Rotator A wheel-like device for slowly (ca. 1 rpm] rotating and gently agitating cultures, usually in a vertical plane.

Runaway Replication Uncontrolled replication of plasmid molecules which multiply within the host building up to several thousand copies. This usually halts cells division but is a very powerful way of amplifying a gene product Plasmid vectors which are temperature sensitive for a copy number control element have been constructed as runaway vectors.

S

S Gene Complex Group of genes in certain plant families (e.g., Cruciferae) that prevent self-fertilization. Thus pollen tubes are inhibited from growth by ovules carrying identical alleles of such genes.

S Phase The cell cycle phase during which DNA synthesis occurs.

Saccharification The Hydrolysis of a polysaccharide to produce a sugar solution.

Saccharomyces (Yeasts) A genus of unicellular ascomycete fungi that can live in both aerobic and anaerobic conditions. They are important in the brewing and banking industries respectively for the alcohol and carbon dioxide they produce by anaerobic respiration. Reproduction is generally asexual by budding although in adverse conditions sexual spores may be formed.

Sacropyte An organism which obtains its nutrients in the form of organic matter in solution from dead or decaying plant or animal tissue.

Salt Tolerance The ability to tolerate or resist 9withstand) a concentration of sodium (Na+ ion) in the soil (or in culture) which is damaging or lethal to other plants. Sodium toxicity is a global problem on irrigated lands. Breeding and screening for increased tolerance in agronomic plants is of great current interest.

Saltatory Replication Increase in the number of gene copies of a specific DNA sequence by laying down numerous strands of DNA in parallel without nuclear division.

Sample (Random) A finite series of observations or individuals taken at random from the hypothetical infinitely large population of potential observations or individuals.

Sampling Error Deviation of a sample value from that of the population due to the small size of the sample.

Sanitizer Agent that reduces microbial numbers to safe levels according to public health requirements.

Saprophyte An organism (usually a bacterium or fungus) that obtains its carbon and energy directly from dead organic matter. During the period of their obtaining food for themselves, they gradually break complex organic compounds like animal dung, rottenwood, stagnant water into simpler compounds. This process of distruction is done after secreting certain enzymes.

Sarcoma A solid tumor growing from derivatives of tissue such as embryonal connective tissue, bone, muscle and fat.

Satellite DNA Eucaryotic DNA that bands at a different density than that of most cellular DNA upon equilibrium centrifugation. In many cases it consists of highly repetitive DNA: in other cases it arises from organelles.

Satellite A short end segment of a chromosome that is separated from the rest of the chromosome by a secondary constriction that is often of thread-like appearance and of somewhat variable length. In certain cases the diameter of the satellite is smaller than in the other parts of the same chromosome.

Sausage Fermentation A bacterial fermantion catalyzed by species of Lactobacillus, leuconstoc and Pediococcs, as well as strains of Micrococcus (which reduce nitrate to nitrite) aiding preservation of these meat products.

Scaffolding A technique which enables any gene or DNA sequence to be integrated into the genome of a desired organism. It was originally described for Bacillus subtilis but is applicable to most organisms.

Scal-Up The development of a full-scale production facility from an initial laboratory concept or experiment.

Schizophyllaceae Basidiocarp: fan-shaped or lobed; basidiospores; colourless and smooth. The organisms are characteristically lignicolous. Genera; e.g. Schizophyllu. Sparassidaceae. Basidiocarp: branching, flattened monomitic lobes, the hymenimum being on both sides of the erect lobes. Genera e.g. Sparassis.

Sclerenchyma A plant tissue, characterized by thick walls. The main function of schelerenchyma is to provide mechanical support.

Scleroprotein One of a group of proteins obtained from the exoskeletal structures of animals. This group exhibits a wide range of both physical and chemical properties. Typical examples of scleroproteins are keratin (hair), elastin (elastic tissue), and colleagen (connective tissue).

Screening A technique for mutant enrichment, e.g. in penicillin screening the utagenised culture is placed under conditions in which wild-type, but not mutant, bacteria will grow. This culture is then treated with the antibiotic penicillin, which kills the growing (wild-type) cells but has not effect on the non-growing (mutant) cells. The population of viable bacteria is thus enriched for mutant cells.

Secondary Association The side by side association of bivalents in meiosis. As indication to the polyploid nature of the organism.

Secondary nondisjunction The phemonenon which results in gametes with extra chromosomes owing to the abnormal karyotype of the original germ cell and not to a a nondisjunctional event during gamete formation.

Secondry nucleus The nucleus resulting from the fusion of the two polar nuclei in the embryo-sac.

Secondary Parasitoid An insect parasite of a primary parasitoid; of, Hyperparasitoid, Primary Parasitoid.

Secondary Protein Structure The helical or extended structure of a polypeptide chain.

Secondary Structure Those aspects of the structure of macromolecules that depend on hydrogen bonding. These include the helical arrangement found in DNA and the alpha-helix or beta-pleated sheet structures found in proteins.

Secretion In the kidney, addition of a material to the urine after the glomerular filtration step.

Secretor Gene Dominant autosomal gene in the human that allows secretion of A and B type blood group antigen into saliva and other body fluids.

Sectorial Chimera A chimera in which the tissues of different genotype appear in distinct sectors.

Seed-Stock Seed which is used by the seeds man to sow for the immediate production of commercial seed.

Selection Coefficient The measure of reduced fitness of a given genotype; it equals one minus the adaptive value of that genotype.

Selection Culture Utilizes difference (s) in environmental conditions or more usually in culture medium composition, such that preferred variant cells or cell lines (presumptive or putative mutants) are favoured over other variants or the wild-type.

Selection Differential In artificial selection, the difference in mean phenotypic value between the individuals selected as parents of the following generation and the whole population.

Selective Advantage That genotypic condition of a cell or individual or genetic class of individuals which increases its chances, relative to others, of representation in later generations of cells or individuals.

Selective Mating Nonrandom mating. Phenotypic differences influence the selection of a mate.

Selective Medium A growth medium containing a component that will inhibit the growth of certain microorganisms an either enhance or not affect the growth of other organisms.

Self-antigen A compound produced by an organisms that stimulates its own autommune system, resulting in damage to its body.

Self-fertilization (Autogamy) Seed set after spontaneous self-pollination occurs in several species of higher plants 9wheat, oats, barely, tomatoes, pease, beans, etc.) and may be enforced in inbreeding experiments with cross-fertilizing plants.

Self-sterility Gene Gene that operates agains itself-fertilization in monoecious plants, usually by control of the rte of pollen tube growth down the style.

Semi-Heterotypic Division A first division of meiosis which gives rise to a Restitution Nucleus following defective pairing.

Semicontinuous Culture The maintenance of cells in a culture vessel in an actively dividing state by periodically draining the medium and adding fresh medium.

Semipeermeable Membrane A poprous membrane that permits the passage of small molecules such as water and inorganic ions, but does not permit the passage of larger molecule such as sugars and proteins. Most biological membranes are semipermeable.

Semisynthetic Descriptive of a compound that has been derived by the chemical modification of a biological product. Examples include some penicillins and steroids.

Sensitization The increase in the reaction of an organism or cell to an antigen artificially induced, e.g. following vaccination.

Sepharose A proprietar form of agarose supplied 9n bead form for use in gel filtration and other types of liquid chromatography. Sepharose is a registered trademark of Pharmacia Biotechnology AB.

Septate Hypha Hypha with cross walls, or septa, which divide each filament into individual cells with individual nuclei.

Sequestration The process of complexing metal ions, usually to chelating agents, to render them ineffective.

Serial Section Series of consecutive tissue sections, usually made with a microtome.

Sewage Farming Sewage farming, an alternative method of sewage treatment is in effect the use of sewage for irrigation. It is a method which maybe employed in areas of low rainfall. This method of treatment may be seen in Melbourne, Australia, where the sewage farm at Werribee, about thirty kilometers from the city centre. The Werribee farm has served nearly two million people; 520,000 cubic metres per day of sewage has been disposed of the 108 square kilometers of land. The precise procedure used (land filtration, grass filtration or lagooning) has depended on the season. While the nutrient load has been reduced by about 50 percent, the run-off passes into Port Phillip Bay contributing a significant nutrient load. The existing sewage treatment facilities have been overloaded, but there will be adequate alternative and additional treatment facilities during the latter part of the decade when the southeastern Purification Plant at Carrum comes into operation.

Sewage Fungus Unsightly slimes resulting sometimes from the pollution of rivers by discharges or organic matter.

Sewerage System The system that collects and carries sewage from source to point of treatment and disposal.

Sex determination The process by which a spore (haploid) or egg 9haploid or diploid) comes to develop the properties of cone or other sex.

Sex Linkage The pattern of inheritance characteristic of genes located on the se chromosomes of organisms having a chromosomal mechanism of sex determination. The sex that is diploid with respect to sex chromosomes can assume three genotpes; homozygous wild type homozygous mutant, or heterozygous carrier. The other sex, haploid for sex chromosomes, is either hemizygous wild-type or hemizygus mutant.

Sex Reversal A change in the sexual character of an individual from a male to female or vice versa.

Sex-conditioned Character Phenotypic character that, although determined by an autosomal gene, is conditioned by the sex of the individual.

Sex-Influenced Traits The difference in hormonal systems between male and female individuals may cause differences in gene expression (e.g., the presence of beards in male humans and the absence of this trait in females).

Sex-linked Gene A gene or locus located in the X chromosome in the species having XY type of sex mechanism with Y chromosome being almost inner. The gene in X whether dominant or recessive finds expression in the phenotype since there is no corresponding gene in Y to modify the action of the gene in X.

Sexual Reproduction Reproduction by the production of male and female gametes, followed by fertilization.

Sickle Cell Anemia A condition produced by an abnormal hemoglobin molecule in the homozygous condition, h^8/h^8. Red blood cells become sickle shaped under reduced oxygen tension.

Sickle Cell Trait Trait Shown by individuals characterized as heterozygote for the sickle cell gene h^8/h^8. All of red blood cells will become sickle-shaped if oxygen is reduced sufficiently. More reduction is tolerated than by an h^8/h^8 individual.

Sieve Tube A column of cells formed form sieve elements, in which food is translocated in plants.

Sigma Factor A subunit of RNA polymerase, which recognizes specific sites on DNA for the initiation of RNA synthesis.

Single Cell Line or Cell Strain A culture started from a single plant cell, usually from suspension cultures of single cells and small aggregates plated on solidified medium. The latter may in corporate a selective agent, from which tolerant or resistant individual cell lines or cell clones a can be selected.

Single-factor Inheritance The control of one character by one gene. This gives rise to discontinuous variation in such characters and intermediates between the dominant and recessive forms of the gene do not usually occur.

Site-specific Mutagenesis, Site-directed mutagenesis, Directed Mutagenesis A Technique in which a cloned gene is specifically mutated in vitro and then used to replace the wild-type copy of the gene in the donor organism. The technique has two advantages over conventional mutagenesis procedures; (i) The proportion of desired mutants obtained is high, up to 50% of the viable cells are mutant in some methods. (ii) No other genes in the organism are mutated and therefore all the desirable characteristics of the stain are retained.

Site-specific Recombination Recombination of short, specific nucleotide sequences. It does not depend on DNA homology, Compare General Recombinations.

Slant Culture A method used for maintaining stock cultures of a microorganism. The cells are placed on a jar contained in a sealed bottle. The bottle or tube is held at an angle while the gel sets.

Slant Tube Fermenter A method used for maintaining stock cultures of a microorganism. The cells are placed on a jar contained in a sealed bottle. The bottle or tube is held at an angle while the gel sets.

Smear A thin layer of material, e.g. bacterial culture, spread on a glass slide for microscopic examination. Also referred to as a film.

Smooth Muscle Involuntary, non-striated muscle cells which are found in the lining of internal organs (e.g., gizzard) and blood vessels.

Smuts Basidiomycete fungi of the order Ustilaginales. Many are important parasites of cereals and form a mass of sooty black spores in place of the grain. Examples of smuts are *Tilletia* and *Ustilago*.

Sodium Azide An antimicrobial agent. It is used at 0.2 percent to prevent infection of chromatography columns and gel suspensions.

Sodium Pump An active transport system within a nerve cell membrane, by which sodium ions entering the cell are transported back into the extra cellular medium against a concentration gradient. It maintains the differential concentration of ions either side of the membrane (the outside being kept positive) necessary to produce a resting potential. A carrier molecule, probably a protein, sis thought to be involved. Energy required for the process is derived from ATP.

Somatic Embryo or Embryoid An organized embryonic structure morphologically similar to a zygotic embryo but initiated from somatic (non-zygotic) cells. These develop into plantlets in Vitro through developmental processes that are similar to those of zygotic embryos.

Somatic Hybrid A cell or plant product of somatic cell fusion; as the result of cell or protoplast fusion and implying genomic integration. The process is somatic hybridization.

Spacer DNA DNA that is located between the regions containing the information used to determine the actualamino acid sequence in proteins. These regions may or may not be transcribed by RNA polymerase.

Spike Culture The aseptic culture of a grain inflorescence for micropore development.

Spindle Fibres Fibres which appears to connect the centromere to the poles of the spindle during cell division these are possibly artifacts of fixation which correspond positionally with some latent 'structure' of the spindle cytoplasm but which do not exist as definite until coagulated by chemical reagents.

Stab Culture Culture of bacteria made by stabbing a long needle, carrying some of the bacteria from another culture, deep into an elongated test-tube containing a solid nutrient agar. It represents an easy way of culturing relatively non-fastidious anaerobic bacteria, since the oxygen concentration deep in the agar is low.

Start-up The first stage of operation of a continuous fermentation or anaerobic digetion process, during which the process is brought to steady running conditions.

Stearic Acid (Octadecoic Acid) A saturated carboxylic acid, which is widely distributed in nature as the glyceride ester. It is present in most fats and oilds of animal and vegetable origin, particularly the so-called hard fats, i.e. those of higher melting point.

Sterile Transfer The process of aseptically explanting or sub-culturing: moving plant material from one location or culture to another under aseptic conditions.

Sterile Transfer The process of aseptically explanting or subculturing; moving plant material from one location or culture to another under aseptic conditions.

Strobilus The cone, or characteristic multiple fruit, of the pine and other gymnosperms. Also, a cone-shaped mass of sporophylls found in club mosses.

Subinert A chromosome with a slight genetic effect is said to be subinert.

Subline A cell line regenerated from a unique cell line of a hybrid callus colony.

Submerged Culture A technique for culturing molds and other microorganisms in which the microbes are distributed through the culture fluid by vigorous mixing. It is in contrast to surface culture where the organism grows on the upper surface of he culture fluid.

Submetacentric Chromosome A chromosome with a centromere some-what nearer to one end than to the other.

Succinic Acid A dicarboxylic acid formed by fermentation of sugars. It occurs in alga, lichens, sugars, beets, and other plant substances. Succinate ion plays on important role in the Krebs cycle.

Subvital À gene or genotype that, when expressed, reduces the viability of its bearers significantly below that of the "normal" or standard type but has a viability at least half as great.

Sulfur Cycle A geochemical cycle by which sulfur is utilized and distributed trough the biosphere.

Sulphur Bactria Filamentous autotrophic chemosynthetic bacteria of the order Beggiatoales. They derive energy by oxidaton of sulphides, and build up carbohydrates from carbondioxide. An example is Beggiatoa.

Supernormal Stimulus Any stimulus, or any intensity of a variable stimulus, that is preferred by animals over the natural sign stimulus.

Superoxide Dismutase An enzyme that catalyzes the reduction of the superoxide radical to hydrogen peroxide.

Surface Sterilization The removal of plant surface microflora prior to aseptic excision of explants. Surface sterilization is accomplished by immersion of tissue in one of many strilants, such as calcium hypochlorite or sodium hypochlorite, hydrogen peroxide, mercuric chloride, silver nitrate or bromine water for an empirically determined period of time. Sometimes a preceding ethanol dip or spray is useful, as is the addition of a surfactant to the sterility, agitation or vacuum may improve the efficiency of sterilization. Through washing in sterile distilled water usually follows, to remove the sterility, although in some cases this is not necessary (hydrogen peroxide).

SV40 Virus A small DNA-containing virus of the group papova virus. The name is in abbreviation for Simain virus 40, which reflects the fact that the virus was first isolated from cell cultures of tissues derived from an African green monkey.

Sympathetic Nervous System A division of the autonomic (involuntary) nervous system. Its activities include increasing blood pressure and acceleration of the heartbeat. The neurotransmitter at the sympathetic ferminal is epinephrine or norepinephrine. (contrast with parasympathetic nervous system).

Synapsis The highly specific parallel alignment (pairing) of homologous chromosomes during the first division of meiosis.

Synaptic Vesicle A membrane-bounded vesicle, containing neurotransmitter, which is produced in and discharged by the presynaptic neuron.

Synthetic Plasmid A circular DNA molecule that is constructed from a replican and one of more gene DNA sequences. The synthetic plasmids have a genetic make-up that differs from wild type-plasmids. They are used as cloning vehicles.

T

T-DNA, Transferred DNA The segment of DNA from a Ti or Ri plasmid which is transferred from Agrobocterium to the genome of its plant host and causes tumour formation or root induction.

T-DNA T DNA is stably inherited by plants regenerated from Ti or Ri plasmid induced callus. Foreigen DNA inserted in to the T DNA can also be stably inherited by such a regenerated plant.

T Phages Group of bacteriophages of some what diverse form, but all having double stranded DNA genomes. Anicosahedra head, and a hollow tail section.

T-system A set of transverse tubes that penetrates ekeletal muscle fibers and terminates in the sarcoplasmic reticulum. The T-system transmits impulses to the sacs, which then release Ca^{2+} to initiate muscle contraction.

Tachphylaxis The process by which subsequent doses of a pharmacological reagent produce decreasing effects.

Tail Refers to the helical component, a virus with banal symmetry.

Tandem Satellites Two short segments of chromosome separated from each by a constriction and from the main body of the chromosome by a second constriction.

Target Organ Indicates the location of the initial infection by viruses, particularly animal viruses. After getting established, the infecting particles may subsequently spread to other regions in the body.

Target Theory Hypothesis developed to account for the mutagenic effects of ionizing radiation on cells. It is cell and that one or more ionizing events or 'hits' are necessary within the target area for radiation to produce effectively a mutation.

Taxonomic Key A guide to the identification of an unknown specimen. The most useful keys are dichotomous.

Teichoic Acids They are mainly presented in Cram-positive bacteria and are substituted polysaccharides and consist of ribotol or glycerol residues joined through diphosphoster linkages.

Telocentric A chromosome or a chromatid having a terminal centromere. A condition arising by Misdivision or by X-ray breakage.

Temperate Bacteriophage A bacteriophage capable of replicaing in step with is host bacterium, being thus transmitted through cell division without necessarily causing lysis.

Temperate Virus Lysogenic virus in which genome is integrated or associated with bacterial genome.

Temperature Sensitive Gene A gene which is not expressed when the temperature to which it is exposed falls outside certain specific limits (e.g., 25-30°C). Temperature sensitive genes have been used in gene engineering to improve the production of unusual secondary products that may impair the growth of the host cells.

Template In biochemistry, a molecule or surface upon which another molecule is synthesized in complementary fashion, as in he replication of DNA. In the brain a pattern that responds to a normal in put but not to incorrect inputs.

Terminal Affinity The property by which chromosomes are held together end to end form Diplotene till first Metaphase of Meiosis (by a Terminal Chiasma) or are brought together in this way at Metaphase.

Terminal Chiasmata Association of homologous chromosomes at meiosis, involving only the chromosome ends. This may be the result of a crossing over event at the chromosomal extremity, or of terminalization of chiasmata which earlier were positioned some distance away from the chromosome terminus.

Terminalisation The process in which the chiasmata move (Terminalize) towards the chromosome ends, the chiasma number thereby being reduced. This phenomenon occurs during the latter part of the prophase of meiosis.

Termination Codons The three codons UAA, UAG, and UGA, which signal termination of a polypeptide chain.

Termination Sequence A DNA sequence that appears at the end of transcriptional unit and signals the end of transcription.

Terpenoid Any of a large number of compounds with varying numbers of carbon atoms, but which are clearly derived from five carbon precursors. Terpenoid quinines and chromonols include phylloquionone (vitamin K), plastoquinones, ubiquinones and tocopherols. Many are important intermediates in electron transport systems.

Test Tube Baby Production of a child following in vitro fertilization, after which the embryo is transplanted into the uterus of its biological mother or a surrogate mother to permit implantation and normal gestation.

Test Tube Fertilization Pollination, followed by fertilization occurring *in vitro*.

Tetracycline Resistance The ability to survive treatment with the antibiotic tetracycline.

Tetracyclines Group of broad spectrum antibiotics obtained from various species of the mould streptomyces. They bind to the 30S subnit of the bacterialribosome, preventing attachment of the aminoacyl tRNA.

Tetrad A group of four cells that in plant are formed as the result of meiosis in a spore mother cell. In flowering plants the tetrads on the male side consist of four microspores, which later separate from each other and develop into pollen grains. In the zoological literature the tetrad is sometimes used to designate a bivalent that is composed f four chromatids.

Tetrazolium Salts Complex organic molecules that changes colour on mild reduction with the formation of insoluble formazans.

All such compounds are characterized by a heterocyclic ring structure containing one carbon and four nitrogen atoms, one of which is quaternary. Two types of salts exist diteraczoles with two such rings. Tetrazolium salts are used as qualitative markers of oxidative enzyme activity, and in tissue sections, this enables enzyme activity to be localized.

Thalassemia An inherited anemia that shows codominance an individual heterozygous for the trait has thalassemia minor, while the homozygote exhibits the severe form, thalassemia major.

Thallophyte A plant having notrue stem, toots, or leaves includes the algae and fungi.

Thelyotoky A type of parthenogenetic reproduction in which only female progeny are produced cf. Arrhenotoky Deuerotoky.

Thermic Shock Exposed to a reduced temperature for several days, and allegedly beneficial pretreatment for immature flower buds or anthers prior to anther culture or pollen culture.

Thixotropy The property, of liquefying when agitated and of returning to the 'gel' condition on standing. This property is shown by smecolloids.

Three Breed Crossing A system of breeding in which breed a is crossed with B and the F1 with C, this hybrid is then crossed with B and the product is crossed with A to produce a hybrid which is crossed with C the next cross is with B and then A and so on.

Thrombin An enzyme that converts the soluble protein fibrinogen into the fibrous fibrin during blood clotting. It is formed from prothrombin under the influence of thromobplastin, calciumions, and other factors, which are activated when blood is removed from the circulation, usually by injury.

Thylakoid A flattened sac within a chloroplast. The membranes of the numerous thylakoids contain all of the chlorophyll in a plant, in addition to the electron carries of photophosphorylation. Thelakoids stckto form grana.

Ti Plasmid A portion of a genome of the bacterium Agrobacterium tumefaciens, the agent involved in crown gall disease. This plasmid is a useful {experimental} vector for the transfer of genetic information into plant cells.

Tissue Culture The growth of cells, tissue, or organs in suitable media in vitro. Such media must normally be sterile, correctly pH balanced, and contain all the necessary micro and macronutrients, carbohydrates, vitamins, and hormones for growth.

Tissue Typing Use of serological tests to determine the major his to compatibility complex antigens of potential transplant donors or recipients.

Tissue A group of cells that is specialized for the particular function. Examples are connective tissue, muscular tissue, and nervous tissue. Several different tissues re often incorporated in the structure of each organ of the body.

Total Pollution Loading A concept which regards the aggregate mass of a specific pollutant, discharged to the environment in a given time period, as more important than the concentration of a pollutant as a basis of environment protection legislation. The total loading on anyone segment of the environment becomes the basis of control.

Trans Acting Locus Gene locus that affects another gene locus on another chromosome, generally through the production of regulatory protein Compare Cis Acting Locus.

Transcription or genetic Transcription DNA dependent RNA synthesis. The process by which cellular RNA molecules are synthesized as determined by homologous NA sections where their sequence recurs in complementary form {the template or transcription strand}.

Transcription the process whereby the message carried in the DNA is transcribed to the RNA through complementary base pairing of the RNA nucleotides with the template DNA. The resulting RNA carries the message to make the specific protein coded in the DNA segment, and it is called messenger DNA (mRNA).

Trans-differentiation Change in the state of differentiation of a cell or tissue. It is phenomenon of relatively infrequent occurrence, but is known to occur in Hydra, and at least in experimental situations in plants.

Transfer RNA (tRNA) Any of at least twenty structurally similar species of RNA, all of which have a MW-25,000. Each species of tRNA molecule is able combine covalenty with a specific amino acid and to hydrogen bond with at lest one mRNA nucletode triplet. Also called adaptor RNA.

Transformant Cell or organism that has been transformed by exposure to foreign DNA.

Transgenation A heritable change taking place at a single gene locus, presumably caused by chemical alteration of the individual gene; a mutation proper, point-mutation or genovariation.

Transgressive It is a phenomenon in which some individuals in the progeny show characters exceeding in value than those shown by either of the parents. The transgressive inheritance is usually seen in crosses involving quantitative characters.

Transition A mutation caused by substitution of one purine for another or one pyrimidine for another.

Transketolase An enzyme of carbohydrate metabolism that catalyzes the transfer of a two crbon ketol moley from a keto sugar to an aldo sugar.

Translational Amplification Use of the same mRNA molecule many times allowing the synthesis of large amounts of polypeptide from a single mRNA and therefore from a singe gene product.

Transplant Shock Refers to the stress involved when Stage II or Stage ill cultures are transplanted to soil (Stage IV) many or all of he regenerated shoots or plantlets may die if suitable care is not taken to acclimatize them gradually to the soil environment particularly prevention of water stress.

Transport (Across bacterial Membranes) Only small, nonionic molecules 9especially lipophilic ones) can diffuse passively across the hydrophobic barrier of the cell membrane; such molecules includes oxygen, carbon dioxide, ethnol, and certain amino acids. Most substances therefore enter or leave the cell as a result of specific transport mechanisms which may or may not require metabolic energy, and which are thought to involve specific protein 'carriers located within the cell membrane.

Transposon A genetic element that can be duplicated and moved to other sites on the same or different genome.

Trap Crop A small planting of a susceptible and highly attractive host, planted early in the season, or removed in space from the main crop, in order to divert attack and infestation by pests and allow for their easy destruction.

Trichlogyne Specialized hypha that extends form the fruiting body of some fungi and forms a receptive organ for the male gamete.

Trickling Filter A secondary treatment process where sewage is trickled over a bed of rocks so that bacteria can break down organic wastes.

Trihyrid Cross A cross between individuals that have different alleles at three gene loci.

Triplicate Genes Three genes, anyone of which produces the same phenotyple effect, e.g., seed coat colour in wheat may be produced by R_1, R_2, R_3.

Tris Borte Buffer A buffer consisting of Tris base and boric acid. It is one of the common buffers for the electrophoresis of DNA and RNA through agarose and polyacrylamide gels. It consists of 89 mM tris base, 89 mM boric acid, and 2mM EDTA (pH 8.0).

Trisomic An individual showing an extra romosome in addition to the normal chromosome set expressed by $2n+1$.

Trition Detergent Uncharged (non ionic) detergent widely used in cell biology for solubilizing the proteins of cell membranes. The commonest is Triton x 100.

Trivalent A group of homologous chromosomes paired with each other along certain chromosome segments. Trivalents may occur from zygotene to first metaphase of meiosis and are best studied at diakinesis and first metaphase.

tRNA Gene Gene that codes for transfer RNA. The total number of these genes in eh genome varies from 360 in yeast to over 1000 in humans. Each of the 20 common amino acids have specific tRNA genes coding for molecules with specificity of binding to their coding triple on the message and to their own structure.

tRNA Isoacceptors Differing species of tRNA that bind the same amino acid but, because of coding degeneracy, have differing Anticodon triplets.

Trophic Nucleus The larger of two nucleiin a cell: vegetative nucleus; macronucleus; trophonucleus.

Trophoplasm 'Vegetative' or nutrititive protoplasm. Unmodified cytoplasm as distinct form kinoplasm.

Trophozoite The vegetative cell of a protozonan, i.e., active feeding stage.

True Breeding Production of progeny which like their parents, express a particular Genotypein their Phenotype.

True to Type Applied to a plant or propagation source this term denotes correct cultivar identification and lack of variation in productivity or performance.

Trypan Red An acidic, red diazo dye, it is used in the early days of chemotherapy for the treatment of trypanosomiasis.

Tube Nucleus The nucleus which occurs in any growing pollen tube and which does not take part in fertilization but is concerned with regulating the growth and behaviour of the tube.

Tubulin Blobular protein suunits (MW'2 55,000 and 57,000 and 57,000) whose regular helical packing forms the hollow, cylindrical microtubules.

Tumble Tube A glass tube closed at both ends with a side neck opening which is commonly attached to a slowly revolving platform. Suspension cultures are agitated and aerated as medium flows form one end of the tube to the other as it is inverted.

Tumor Mass of tissue resulting from uncontrolled growth

Turbidimetry A method of estimating bacterial growth or populations by measurement of the degree of opacity (or turbidity) of the suspension.

Twisting Number Number of base pairs in a DNA duplex sequence, divided by the number of base pairs for every helical turn.

Two Point Cross Genetic cross in which either one parent is a double mutant (--) and the other is wild type (+ +) with respect to two markers being followed in the cross (an 'in coupling' cross) or each parent carries one of the two markers (+ - X-+) being followed in the cross (an 'in repulsion' cross).

Two Step Digester An anaerobic digester in which the acidogenic (acid forming) reactions are from the methanogenic reactions.

Tyndallization Fractional sterilization; exposure to live steam for 30 minutes each day for three successive days to kill vegetative cells.

Type Culture A particular species considered representative of the characteristics of the species and used as a reference culture.

U

U Protein Hypothetical protein thought to regulate the transition of cells from G0 to G1 phase of the cell cycle, and thus inevitably into *S* phase. The idea would be that the concentration of this untable (*U*) protein would have to exceed a threshold level for triggering progression through the cycle, and that this would only happen if the cell has adequate access to growth factors or to nutrients, Also known as Trigger Protein.

U1 RNA U1 RNA thought to act as an external guide sequence to facilitate the correct splicing of introns from mRNA precursor molecules. It contains sequence at its 5' end which are complementary to the conserved intron sequences near the intron/ exon junction.

UAA A codon that does not code for any amino acid. It is termed a nonsense codon and its functions is to stop translation.

UAS, Upstream Activator Sequence A part of a eukryotic promoter some distance away from both the transcriptional start site and the tata box. The UAS affects the extent of transcription from a given promoter without affecting the site of transcriptional initiation.

Ubiquitin Acidic protein widely, and perhaps universally, distributed among eukaryotes. It binds strongy to chrmatin and may have a structural role in chromatin conformation. Bindings specifically to hitone H2A by a covalent linkage, the complex has come to be known as protein A24.

Ulcer A rounded or irregularly shaped area of inflammatory tissue destruction in the epithelial lining of a surface.

Ultra-contrifugation (Prefix ultra plus Latin centrum, centre and Latin fugere, to flee away from). Refers here to the process of sedimentation of suspended particles subjected to very high centrifugal force in a centrifuge. Since different kinds of particles sediment at different rates, the process offers a convenient means for ascertaining the homogeneity of a suspension. Moleclar weights and sizes of the seddimenting particles can also be determined from the rates of sedimentations.

Ultra-structure Latin prefix ultra, extreme plus Latin structural, to build by assembling components). Refers to the extremely detailed analyses of the structure of a cell, its organelles and other components. Involves sophisticated techniques, such as X-ray crystallography and electron microscopy.

Ultrafiltrton Fermenter A continuous fermentation system run in conjunction with continuous ultrafiltration, such that the low molecular weight product of the reaction is removed, thus reducing the effect of feedback inhibition.

Ultramicrosophic Visible with the electron microscope; too small to be seen with the conventional light microscope.

Ultrasonic Cleaner A device to include high frequency vibration of materials, removing adhering substance from surfaces by mechanical action. This device is useful of or cleaning glassware and for disinfecting plant material.

Ultrasonic Waves High frequency sound waves, used for destruction of bacterial cells.

Ultraviolet Absorption Curve Plot that shows the relationship between differing wavelengths of light in the ultraviolet range and the percentage absorbance by a solution of biological molecules, Such a curve helps to distinguish nucleic acids from proteins; the former absorbs maximally at about 260 nm and the latter at about 280 nm.

Ultraviolet Disinfection The destruction of microorganisms in close proximity to a ultraviolet light source of 253 nm. The technique is used in contaminated environments and clean rooms.

Ultraviolet Irradiation Irradiation with ultraviolet light, which may induce melanin formation in low doses, but is also a mutagen and, in high doses, lethal to living cells. It is widely used for the maintenance of sterility in cabinets used for tissue culture, but will not penetrate glass or plastic efficiently. Its mutagenic effect results from its injunction of pyrimidine dimmers in DNA.

Ultraviolet Microscope Microscope in which ultraviolet light is generated from the light source, usually a high-pressure mercury vapourlamp, and which is provided with excitation and barrier filters so that ultraviolet light can excite fluorescent molecules in the specimen, harmful ultraviolet light can be filtered out, and fluorescent light only allowed to pass through the eyepiece.

Underlapping Possible outcome of collision between two cells in culture, particularly head-side collision: one cell crawls underneath the other, retaining contact with the substratum, and obtaining traction from contact with the rigid substratum (unlike overlapping, where traction must be gained on the dorsal surface of the other cell.)

Unequal Bivalent A bivalent in which one of the constituent chromosomes is longer than the other, so that it has an unpaired region at one end.

Unequal Crossing-over Imprecise pairing of duplicated segments prior to crossing-over, leading to chromatids with unequal copies or numbers of copies of the segments concerned.

Univalent A body at the first meiotic division corresponding with a single chromosome in the mitotic complement, especially when unpaired bivalent, trivalent, quadrivalent, quinquevalent, sexivalent, septivalent, octavalent, etc. are associations of chromosomes held together between diplotene and metaphase of the first division of meiosis usually by chiasmata. Trivalents and high association are collectively termed multivalents.

Unmatched S Gene Any S gene which is represented either in the style or pollen but not in both, in given pollination (An s gene is any gene controlling he specificity of incompatibility reactions in plants).

Unscheduled DNA Synthesis Synthesis that occurs at times other than the S phase of the cell cycle. Most of such synthesis is DNA repair, although synthesis of mitochondrial DNA in the cytoplasm is also continuous.

Untranslated Sequence A region of mRNA that is not used in the synthesis of anamino acid sequence of a given peptide or protein. They usually appear at either end of the sequence that codes for the amino acid sequence.

Unwinding Protein Protein that untwists the DNA double helix before the replicating fork reaches that region during DNA synthesis. Such a protein, discovered in phage T4, has been termed the gene 32 protein.

Upflow Sludge Blanket An advanced anaerobic digester in which the cells form flocs which are kept in suspension in the lower part of the reaction tank in the form of a bed. The system is used with a liquid feed for aste materialin solution. The feed streams introduced at the base of the fermenter so that it passes slowly up through the bed.

Uric Acid A compound that serves as the main excreted form of nitrogen in some animals, particularly those which must conserve water, such s birds, insects, and reptiles.

Unridine diphosphate Galctose Nucleoside diphasphate sugar, found as an intermediate in the metabolism of galactose, and formed in a reacton between uridine triphosphate (UTP) and galactose-1-phosphate catalysed by the enzyme UDP-Gal pyrophosphoylase: UTP + alactose-1-phosphate – UDP-Gal + PP1. UDPG epimerase then catalyses an inversion of configuration converting UDP-Gal to urdine diphosphate glucose.

Uronic Acid A member of the most important group of sugar acids sugar acids, the common constituents of polysaccharides. In urnic acids the carbon atom carrying the primary hydroxyl group (the CH_2OH end) is oxidized to a carboxyl group (COOH); as glucuronic and galactruronic acids.

UTP Uridine Triphosphate. The corresponding deoxy derivative does not exist. Hence labeled UTP is used to specifically label RNA.

UV Clothing Discard Rack A rack that holds a standard laundry bag and is protected at the top with a curtain of UV. Clothing worn in he contaminated laboratory is discarded into this laundry bag in the contaminated change room.

V

V Factor Nicotinamide adenine dinucleotide required for the growth of certain *Haemophilus* spp.

Vccination Inoculation with a biologic preparation (a vaccine) to produce immunity.

Vaccine (latin vacca, cow) A suspension of weakened or killed viruses or any other pathogen which when injected into the body of a susceptible host, immunizes the latter, against the same type of pathogen, or even against their toxins. It is a product which contains an antigen, consisting of killed pathogenic organisms, living avirulent (attenuated or weakened) organisms, or inactivated exotoxins (toxoids).

Vacuum Fermentation A fermentation system developed for the continuous production of ethanol and other volatile products. The reactor is run under reduced pressure so that the volatile product leaves the fermenter continuously in the exhaust gas. The method, which is of interest for thermophilic fermentations, is still largely at the experimental stage.

Vagina, Artificial A double walled rubber tube to which is fitted at one end, a glass receptacle and which is used in artificial insemination in collecting semen from the male. The space between the double walls in usually filled with warm water.

Variable Antigen Term usually applied to the surface antigens of those parasitic or pathogenic organisms which can alter their antigenic character to evade host immune responses.

Variable Antigen Term usually applied to the surface antigens of those parasitic or pathogenic organisms which can alter their antigenic character to evade host immune responses.

Vsoactive Amines Products such as histamine and 5-hydroxytryptamine released by basophils, mast cells and platelets which act on the endothelium and smooth muscle of the local vasculature.

Vasotocin Peptide hormone secreted by the posterior pituitary gland of lower invertebrates and by the pineal body in mammals. It increases skin permabilty in amphibins and regulates bladder function and is structurally similar to Vasopressin.

Vegetative Reproduction A sexual preproduction by buds, layering, cutting, etc., with the intervention for formation of gametes and fertilization.

Vegetative Non reproductive, or non flowering, or asexual, the concerned with the growing stage of a microorganism as opposed to the spore state.

Vertical Spread Transfer of infectious agent from parent to offspring vi ovum or sperm or by contact between placenta and offspring.

Viable Literally, "capable of life," and generally refers to the ability of microbial cells to grow and multiply as evidenced by for example, the formation of colonies on an agar culture medium. Frequently organisms maybe viable under one set of culture conditions and not under another set, making it extremely important to define precisely the conditions used for determining viability.

Virulence The relative pathogencity of an organism. Virulence depends on the invasiveness and toxicity of the pathogen. Virulence may very between strains of the same organism.

Vitamin A The fat-soluble vitamin controlling night-blindness and the disease resisting power of mucous membranes. Vitamin A is manufactured by the liver from carotene.

Vitamin B complex A group of ten or more water –soluble vitamins, which tend to occur together.

Vitamin C (ascorbic Acid) A water soluble vitamin, which is widely required in metabolism. The major sources of vitamin C are fresh fruit and vegetables and severe ascorbic deficiency results in scurvy.

Vitamin D-resistant Rickets Sex-linked dominant hereditary disease in humans involving defective renal transport of phosphorus. The disease resembles rickets due to vitamin D deficiency, but persists in the presence of adequate dietary vitamin D.

Vitamin K A fat-soluble quione involved in photosynthetic electron transfer in plants and blood clotting in animals.

Vitamin P A flavone derivative concerned with permeability; citrin.

Von Hippel-Lindau Disease Inherited disease caused by an autosomal dominant gene, characterised by reinal haemangioma and renal carcinoma.

W

Waagen, Mutations of Discontinuous changes of smallor moderate extent; particularly used ofdiffernt forms of fossils of the same lineage but belonging to different horizons.

Waggle Dance The running movement of a working honey been on the hive, during whchthe worker traces out a repeated figure eight. The dance contains elements that transmit to othr bees the locaiton of food.

Walling-off Separation of diseased from healthy tissue by barrier tissues produced by diseased plant.

Waltzer Mutant strainof laboratory mice that carries a neurological defect. Homozygous animals tend to walk in circlesand have a head-shaking behaviour.

Warm Antibodies Most IgG antibodies react better at 37°C than at lower temperatures, especially against red cell antigens. tese are the warm antibodies as contrasted with cold agglutinins, especially IgM, which agglutinate below 28 °C.

Warning Appearance A conspicuous trait, such as bright coloration, servingto warm potential predators that the animal is poisnous or otherwise.

Water Potential A measure of energy for water in a plant. Water potential is a fundamental thermodynamic function such that water moves from a pointof high water potential to one of lower water potential.

Whole Cell Path A vriant of patch clamp technique, in which the patch electrode seals the cell, with direct communication between the interior of the electrode and the cytoplasm.

Wild Type An Individual as it occurs in nature. The wild type is usualy denoted by '+' sing with a supercript (consisting of first letter/letters of the character) to the right of the '+' sign. The wild type allele is usually dominant but it may be recessive also e.g. the wild allele of barmutant in Drosophila.

Wobble Ability of third base in tRNA anticodon (5' end) to hydrogen bond with any of two or three bases at 3' end of codon. Thus, a single tRNA speies can recognize several different codons.

Wright's Inbreeding coefficient (F) Proportion of loci at which an individual organism is homozygous.

X

X Chromosome (1) One of the sex chromosomes. A female is XX in Drosophila and in many mammals, including man. A male is XY. In poultry, the male is xx and the female is XO. (2) with diploid sex differentiation, the sex chromosome in regard to which one sex is homozygous (the Homogametic Sex.) With haploid sex differentiation (Bryophyta, etc.), the sex chromosome of the female.

X Inactivation Repression by condensation of the transcriptional activity of one of the two copies of the *X* chromosome in all somatic cells of the female mammal. One of the two *X* chromosomes is randomly inactivated in early embryonic life, as a mechanism of dosage compensation.

X-radiation Radiation of X-rays, which are produced when metal surfaces are bombarded withhigh-speed electrons. X-RAys are ionizing radiations and have wavelengths intermediate between ultraviolet and gamma-radiation.

X-ray Breakage The following terms are used in describing the results of X-ray breakage. They have alsobeen applied to cases of spontaneous breakage.

X-ray Therapy The treatment of disease, such as cancer, using controlled levels of X-rays.

Xerocleistogamy Self pollination taking place with in flowers whose perianths have been prevented from opening by excessive dryness.

Xerogel A solution of linear molecules whose movement, relative to one another, is restricted by cross-linking or physical intraction, producinga three-dimensional net-work of solvaed plymer chains. a given polymer matrix only forms xerogels in a limited range of solvents in which the individual polymer chains are inherently soluble.

Xylan A polymer of xylose; a component of hemicellulose in plants. Xylants are used in the production of furfural, which is used in the manufacture of plastics.

Y

Y Linkage Sex linkage, in which genes are located on the *Y* chromosome, and are therefore expressed only inmales (holandric inheritance). In mammals the *Y* chromosome carries few, if anygenes, butin some organism (e.g., medaka fish) the *Y* chromosome does carry genes.

Y-linked Inheritance Inheritance conditioned bygenes in a *Y* chromosome Properties conditioned in this way are exclusively transmitted from males to males.

YCp, Yeast Centrometric Plasmid *A S.* cerevisiae cloning vector which contains a cloned centromere sequence. The centromere confers stability on the plasmid but also reduces its number to one or two per cel. At meiosis, a YCp segregates as a mini-chromosome.

Yeast Autolyzate An autolyzate produced using yeast as the starting material, The autolyzates are the result of the action of intercellular enzymes, primarily proteases, on plymeric substances in the cells.

Yeast Vectors Yeast cells used as carriers of plasmids, in which novel genes may be replicated, or a transformed cells used to produce the products of integrated novel genes.

Yolk Nucleus Vitelline body; a cytoplasmic body in the young oocyte, which serves as a centre for the yolk-formatinin many forms. In many cases it is traceabe to the idiozome and its associated structures.

Yolk Inert, nutrient constituent of the ovum; consisting mainlyof proteinand fat. Ova of viviparous animals tend to have very little yolk.

Z

Z DNA Form of DNA in which the helix is screwed ina left-handed rateher than a righ-handed direction. Z DNA occurs in short stretches of DNA in nature. Compare B DNA.

Z Value The degrees fahrenheit required to reduced the thermal death time by 90% or one long value.

Zimmerman Process A sewage treatment process achieving oxidation of organic matrial in a closed vessel under high pressure and temperature.

Zoogonous Viviparous; giving birth direct to live young, as opposed to the production of eggs.

Zoospore An asexula motile spore produced by a zoosporagium. It has one of more flagella. These motile spores may encyst in adverse conditions, or may be the means by which the fungus pentrates a new host.

Zygonema The chromosome thread during the zygotene phase, when the chromosomes have come together in pairs.

Zygospore Sexual spore which results from the fusion of like gametes; characteristic of Zygomycetes.

Zygotic Meiois Meiosis which takes plae in the firstt wo divisions after fertilization, at the beginning of the life cycle of an organisms; initial meiosis; zygotic reduction.

Zymogenous Soil microorganisms, primarily rods and spore formers that readily decompose organic matter.